Lecture Notes in Computer Science 9650

Commenced Publication in 1973
Founding and Former Series Editors:
Gerhard Goos, Juris Hartmanis, and Jan van Leeuwen

More information about this series at http://www.springer.com/series/7410

Michael Chau · G. Alan Wang
Hsinchun Chen (Eds.)

Intelligence and Security Informatics

11th Pacific Asia Workshop, PAISI 2016
Auckland, New Zealand, April 19, 2016
Proceedings

 Springer

Editors
Michael Chau
The University of Hong Kong
Hong Kong
SAR China

Hsinchun Chen
The University of Arizona
Tucson, AZ
USA

G. Alan Wang
Virginia Tech
Blacksburg, VA
USA

ISSN 0302-9743   ISSN 1611-3349   (electronic)
Lecture Notes in Computer Science
ISBN 978-3-319-31862-2   ISBN 978-3-319-31863-9   (eBook)
DOI 10.1007/978-3-319-31863-9

Library of Congress Control Number: 2016934010

LNCS Sublibrary: SL4 – Security and Cryptology

Printed on acid-free paper

This Springer imprint is published by Springer Nature
The registered company is Springer International Publishing AG Switzerland

Preface

Intelligence and security informatics (ISI) is an interdisciplinary research area concerned with the study of the development and use of advanced information technologies and systems for national, international, and societal security-related applications. In the past few years, we have witnessed ISI experiencing tremendous growth and attracting significant interest involving academic researchers in related fields as well as practitioners from both government agencies and industry.

PAISI 2016 was the 11th workshop in the series. In 2006, the first workshop on ISI was held in Singapore in conjunction with PAKDD, with most contributors and participants coming from the Pacific Asian region. The second workshop, PAISI 2007, was held in Chengdu. Following that, the annual PAISI workshop was held in Taipei, Taiwan (2008), Bangkok, Thailand (2009), Hyderabad, India (2010), Beijing, China (2011, 2013), Kuala Lumpur, Malaysia (2012), Tainan, Taiwan (2014), and Ho Chi Minh City, Vietnam (2015).

Building on the success of these ISI meetings, we held PAISI 2016 in conjunction with PAKDD 2016 in Auckland, New Zealand, in April 2016. PAISI 2016 brought together ISI researchers from Pacific Asia and other regions working on a variety of fields and provided a stimulating forum for them to exchange ideas and report on their research progress. This volume of Springer's *Lecture Notes in Computer Science* contains 14 research papers presented at PAISI 2016. It presents a significant view on regional data sets and case studies, covering such topics as criminal network analysis, malware detection, data mining, and text mining.

We wish to express our gratitude to all members of the Workshop Program Committee and additional reviewers who provided high-quality, constructive review comments within a tight schedule. Our special thanks go to the PAKDD 2016 Organizing Committee and workshop chairs. We would also like to acknowledge the excellent cooperation with Springer in the preparation of this volume, as well as the effective conference management software EasyChair. Last but not least, we thank all researchers in the ISI community for their strong and continuous support of the PAISI series and other related research activities.

April 2016

Michael Chau
G. Alan Wang
Hsinchun Chen

Organization

Workshop Co-chairs

Michael Chau   The University of Hong Kong, Hong Kong SAR China
G. Alan Wang   Virginia Tech, USA
Hsinchun Chen   The University of Arizona, USA

Program Committee

Robert Weiping Chang	Central Police University, Taiwan
Vladimir Estivill-Castro	Griffith University, Australia
Uwe Gläesser	Simon Fraser University, Canada
Eul Gyu Im	Hanyang University, South Korea
Da-Yu Kao	Central Police University, Taiwan
Siddharth Kaza	Towson University, USA
Paul W.H. Kwan	University of New England, Australia
Wai Lam	The Chinese University of Hong Kong, Hong Kong SAR China
Mark Last	Ben-Gurion University of the Negev, Israel
Ickjai Lee	James Cook University, Australia
You-Lu Liao	Central Police University, Taiwan
Hsin-Min Lu	National Taiwan University, Taiwan
Jun Luo	The Chinese Academy of Sciences, China
Xin (Robert) Luo	The University of New Mexico, USA
Byron Marshall	Oregon State University, USA
Dorbin Ng	The Chinese University of Hong Kong, Hong Kong SAR China
Symeon Papadopoulos	Information Technologies Institute, Greece
Shaojie Qiao	Southwest Jiaotong University, China
Shrisha Rao	International Institute of Information Technology - Bangalore, India
Srinath Srinivasa	International Institute of Information Technology - Bangalore, India
Aixin Sun	Nanyang Technological University, Singapore
Paul Thompson	Dartmouth College, USA
Jau-Hwang Wang	Central Police University, Taiwan
Jennifer J. Xu	Bentley University, USA
Yilu Zhou	Fordham University, USA

Keynote Speech:
Cloud-Centric Assured Information Sharing

Bhavani Thuraisingham

The University of Texas at Dallas, USA
bxt043000@utdallas.edu

Abstract. This presentation will describe our research and development efforts in assured cloud computing for the Air Force Office of Scientific Research. We have developed a secure cloud computing framework as well as multiple secure cloud query processing systems. Our framework uses Hadoop to store and retrieve large numbers of RDF triples by exploiting the cloud computing paradigm and we have developed a scheme to store RDF data in a Hadoop Distributed File System. We implemented XACML-based policy management and integrated it with our query processing strategies. For secure query processing with relational data we utilized the HIVE framework. More recently we have developed strategies for secure storage and query processing in a hybrid cloud. In particular, we have developed algorithms for query processing wherein user's local computing capability is exploited alongside public cloud services to deliver an efficient and secure data management solution. We have also developed techniques for secure virtualization using the XEN hypervisor to host our cloud data managers as well as an RDF-based policy engine hosted on our cloud computing framework. Finally we have developed a secure social media framework hosted on our secure cloud computing framework. The presentation will discuss our secure cloud computing framework for assured information sharing and discuss the secure social media framework. We will then discuss the relationship to big data security and privacy aspects and connect our research to Secure Internet of Things with a special emphasis on data privacy.

Keywords: Cloud computing • Hadoop • HIVE • Information sharing • Social media • Secure internet of things • Data privacy

Speaker Bio: Dr. Bhavani Thuraisingham is the Louis A. Beecherl, Jr. Distinguished Professor of Computer Science and the Executive Director of the Cyber Security Research and Education Institute (CSI) at The University of Texas at Dallas. She is also a Visiting Senior Research Fellow at Kings College, University of London. She is an elected Fellow of IEEE, the AAAS, the British Computer Society, and the SPDS (Society for Design and Process Science). She received several prestigious award including IEEE Computer Society's 1997 Technical Achievement Award for "outstanding and innovative contributions to secure data management", the 2010 ACM SIGSAC (Association for Computing Machinery, Special Interest Group on Security, Audit and Control) Outstanding Contributions Award for "seminal research contributions and leadership in data and applications security for over 25 years" and the SDPS Transformative Achievement Gold Medal for her contributions to interdisciplinary research. She has unique experience working in the commercial industry (Honeywell), federal research laboratory (MITRE), US government (NSF) and academia and her 35 year career includes research and development, technology transfer, product development, program management, and consulting for the federal government. Her work has resulted in 100+ journal articles, 200+ conference papers, 100+ keynote and featured addresses, eight US patents (three pending) and fifteen books (one pending). She received the prestigious earned higher doctorate degree (DEng) from the University of Bristol England in 2011 for her published work in secure data management since her PhD. She has been a strong advocate for women in computing and has delivered featured addresses at events organized by the CRA-W (Computing Research Association) and SWE (Society for Women Engineers).

Contents

Network-Based Data Analytics

Data and Text Mining

Cybersecurity and Infrastructure Protection

Network-Based Data Analytics

The Use of Reference Graphs in the Entity Resolution of Criminal Networks

David Robinson[(✉)]

Inland Revenue, Wellington, New Zealand
david.robinson@ird.govt.nz

Abstract. Entity resolution (ER) is the detection of duplicated records within a dataset representing the same real-world entity. The importance of ER is amplified within law enforcement as criminal data, or criminal networks, has inherent uncertainty and ER inaccuracy incurs a high cost. Commercial ER solutions focus on fast and scalable resolution of obvious pairs of entities, rather than the more complex non-obvious pairs which are so critical to law enforcement. Here we outline the use of proper names represented as reference graphs - generated from an algorithm that conducts name similarity, logic-based pruning, and classification using community detection and a proper name origin algorithm. The resultant classes are used at indexing and decision management stages within an ER model to support the detection of non-obvious duplicate entities. Utility is clearly demonstrated through the application of the approach on three real-world datasets of varying origin, size, topology, and heterogeneity.

Keywords: Entity resolution · Record linkage · Reference graph · Criminal networks · Indexing · Decision management · Community detection

1 Introduction

Criminal networks - graph representations focusing on criminal actors - present significant challenges in terms of deriving an accurate representation that mimics the real-world. Incompleteness, data heterogeneity, non-intentional error, intentional misinformation, and bias all contribute to increase the uncertainty of the data. At the core of this uncertainty and variance is accurately and reliably resolving duplicate entities within the data representation which in fact represent the same real world entity [1] - entity resolution.

Whether the problem is derived from the integration of multiple heterogeneous datasets or focuses on one homogeneous dataset the domain dictates the nature and complexity of the problem, which in turn places specific demands of an entity resolution solution. This complexity can be driven from artefacts of the source(s) of data and their representation or the wider domain where data error is generated from not only incidental variance but variance derived from specific intent. Interestingly within the criminal domain the very entities that are the source of intentionally poor quality data are often the very entities that are of most interest.

© Springer International Publishing Switzerland 2016
M. Chau et al. (Eds.): PAISI 2016, LNCS 9650, pp. 3–18, 2016.
DOI: 10.1007/978-3-319-31863-9_1

The criminal context provides an additional layer of complexity and uncertainty due to the motivation of entities to actively supply misinformation with the goal to reduce the effectiveness of entity resolution. For this reason entity resolution and link discovery are often deployed in concert to enhance the quality of the graph through making the data as explicit as possible and discover latent knowledge. This paper however is limited to entity resolution, and in particular the use of reference graphs in entity resolution. A critical element though to highlight is that entity resolution in the criminal domain must be able to contend with not just missing nodes and edges, but the existence of fake nodes, nodes that are in the dataset but do not exist in the real world, and spoof nodes, where a real world node will be represented in multiple nodes within the dataset [2]. And of course in instances of high incompleteness, the presence of fake and spoof nodes (and edges), and high uncertainty the generation of false positives can significantly obfuscate the graph.

Therefore, inexpensive, accurate and scalable approaches to entity resolution that go beyond identifying the obvious matches (deduplication) and can also detect the non-obvious matches are of critical importance. Current "state of the art" commercial entity resolution products are often focused on markets that require generic scalable fast deduplication solutions and do not place the requisite emphasis on the detection of the complex low-signal non-obvious matches. Responding to this need the reference graph algorithm has been designed to support the detection of non-obvious duplicate entities.

A reference graph, in this instance, is defined as a graph constructed from a set of proper names whose pairwise distance is calculated using a variety of concepts, including string similarity and co-occurrence, and represented as a graph that can be improved over time to enhance ER performance. Improvement to the reference graph can be derived from improvements in the algorithm that constructs the graph, the integration of additional data, or the manual annotation by human experts. The reference graph algorithm generates meta-data that can be used in both indexing and decision management stages of an entity resolution model that out-performs more traditional algorithms on typical criminal networks, is scalable ($\approx$4 million nodes) and "fast enough".

From an applied perspective two main elements of entity resolution are critical to an effective and performant solution: indexing (otherwise known as blocking or key-generation) and decision management - making a decision on whether to resolve a pair of entities or not. These two components of entity resolution will be briefly introduced to create context required before reference graphs are explained further.

Indexing is essentially the creation of subsets of records or entities based on some notion of similarity. The comparison of all pairs is an intractable problem and hence indexing has been a pragmatic solution to avoid exhaustive comparison and reduce the computational expense by breaking the initial set into multiple sub-sets or blocks. The number, size and the "similarity" between entities within each sub-set determine the quality of the indexing, as each block, or cluster of blocks, will serve as the set that pairwise similarity will be measured. The quality in combination with runtime, scalability, ease of optimization, and versatility (how well the indexing performs across a range of different scenarios) determine the utility of the indexing. Many approaches have been used to generate blocks including the use of phonetic algorithms like Soundex [3],

Double-Metaphone [4], and Metaphone 3 [5] which generate keys based on the phonetic sound of the name (e.g. Metaphone 3 of "Robinson" == "RPNS"). Although many of the latest generation algorithms are proprietary, making them problematic to benchmark against, a wide range of quality algorithms exist and are freely available to apply. Truncation approaches similarly generate keys off a predetermined number of letters from the front of the Family Name (e.g. "Robinson" == "ROB"). Suffix Array is another blocking approach that is used, which creates an integer key based on lexicographically sorted string suffixes [6]. Meta-blocking approaches are another class of approach that takes the output from a blocking strategy and attempts to optimize given the tolerance for error and speed. A good example of a meta-blocking strategy is that outlined by Hernandez and Stolfo [7, 8] and McCallum, Nigam and Ungar [9] using windows or canopies to effectively create overlapping classes to reduce computational expense and yet retain accuracy.

Decision management is about making decisions under uncertainty, given the context of the purpose of making those decisions and validation metrics [10]. The first element within ER to achieve this is the discovery of a number of relevant pairwise similarity metrics (e.g. Name similarity, Date of Birth similarity, and distance), measuring key concepts. The pairwise similarity metrics fall under two concepts, those helping to measure congruence and commonality. Congruence refers to the holistic assessment of how similar the pair is based on name features, other attribute features (e.g. Date of Birth; Gender) and contextual features such (e.g. graph distance; community membership) which have all shown to significantly improve performance [11–13]. Commonality refers to how often these features are represented in different entities within the bounded context of the comparison. If the pairwise similarity is conducted globally across the entire dataset without any notion of distance (for example, geographical or social relationship) between the pair then the commonality measurement has to be based on a global assessment. However, incorporating the notion of distance can create a bounded local context which significantly alters the measurement of commonality. For example, the certainty of the pairwise assessment of whether "Joan Mary SMITH" and "Joan Mary SMITH" are indeed the same real-world entity is significantly increased if it is known that they reside in the same suburb or community detection has identified they are members of the same community. The second component is factoring in the context of what the ER is being conducted for (for example tax evasion detection or counter terrorism). Central concepts from a contextual perspective include how rare the class of events are that are the focus of detection and measurement, and the size of impact, and then how that translates into the cost of false positives and false negatives. The third aspect is the generation, retention, and use of contextual validation metrics of the decision made - a critical element to decision management [14]. In this case the ER model computes transitivity providing a useful guide to the measurement of accuracy of the overall model, each ER function, and at a pair level. Indeed, generating transitivity metrics creates the opportunity for fine-grained transitive closure based approaches to enhance performance [15]. All relevant discovered metrics are then output in graph format enabling robust validation and exploitation of the model.

As alluded to earlier a core feature of the broader entity resolution model developed is the use of reference graphs to drive indexing and support decision management.

Reference graphs are an explicit representation of proper name knowledge derived from the data and from external knowledge sources. Knowledge representations are often used as a "deterministic" adjunct to bolster accuracy through the provision of a set of relationships between names based on synonyms and hypocorisms [16]. Here the novel generation of reference graphs from proper names, derived from both the data source and external proper name based edge-lists of hypocorisms, has been used to generate blocking-keys from a simple partitioning of the reference graphs using non-overlapping community detection coupled with classification derived from a proper name co-occurrence graph, which are used at both indexing and decision management stages. An important feature of the indexing is that the keys form a graph, which creates the opportunity for deploying meta-blocking strategies to optimize performance [17].

The details of how the reference graphs are constructed and implemented are covered before the experimental conditions are outlined and the results thereof are examined. A discussion of the results, conclusions and extensions complete the investigation.

2 Generation and Implementation of Reference Graphs

2.1 Reference Graph Generation

The proper name reference graphs, Family Name Reference Graph (FNRG) and Given Name Reference Graph (GNRG), are generated by measuring the string distance, using the Jaro-Winkler algorithm [18], between all names of the same class, whether that is the Family name class or Given name class. The proper names are sourced from the target unresolved dataset, and potentially any other source. Dependent on the size of the target unresolved dataset there may need to be an intermediate blocking phase to generate the reference graph. The intermediate blocking phase is implemented by a two-step algorithm that firstly blocks the names by the first letter of each name and compares all names beginning with that letter and uses the string distance to start building the reference graph. The second step of the algorithm generates a sample of names and conducts string distance on all pairs from the sample. The complete graph derived from this process is then pruned via using a simple threshold to remove edges and enable the assessment of relationship strength between names starting with a specific letter (see Fig. 1). This derived graph served as the basis in the second stage of the algorithm to select meta-blocks (as per the circled clusters) from which to base additional string distance comparisons.

The result of this intermediate blocking phase is an approximately complete graph based on the distance between proper names, either Family names as in the FNRG or Given names as in the GNRG.

The next step is to turn the distance graph into a similarity graph (so more intuitive) and only retain edges between proper names that are useful to the ultimate goal of conducting community detection that is both accurate and inexpensive. A simple thresholding method could be used to delete edges under a certain weight however that would mean that unique names are more likely to be isolated and not part of larger blocks and therefore resulting in poor blocking performance. So, a core goal is to ensure that communities of names are larger than one or two and obviously at the other end of the

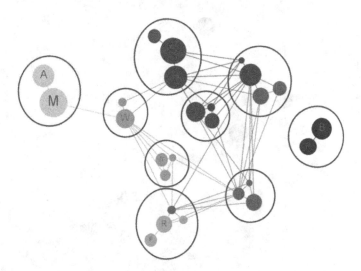

Fig. 1. This figure depicts a derived contracted graph indicating the relationship between classes of names based on their first letter. This graph is used in the second stage of the intermediate blocking phase to improve the accuracy and completeness of the reference graphs.

spectrum are small enough so the communities/blocks enable scalable deployment. To do this the two highest (as the graph is now a similarity based graph rather than a distance based graph) weighted incident edges of every node (i.e. name) is retained to ensure the graph retains a giant single component. It is important to select the two highest as simply using the single highest weighted edge can result in isolated dyads. Using this approach significantly reduces the number of components and ensures the smallest component is a triad. Furthermore, all edges above a pre specified threshold were retained.

An alternate source of names and their variants (derived from transcription, hypocorisms) is then introduced. This secondary source of proper name variation is important from a human centered systems perspective as it creates the opportunity for experts to add their explicit knowledge into the entity resolution model and derive instant performance improvements. Knowledge is added to both create relationships between names and negate relationships between names that do not exist. For example, the transcription of the Chinese family name 韩 into the Latin alphabet is dependent on the dialect - China's pinyin system converts this to "Han", in Cantonese "Hon", and in Hainan "Hang". Furthermore, negation is critical in situations where names are very similar but are actually distinct proper names. String matching cannot discriminate between these sets of names easily, so other methods are required to buttress performance (see Fig. 2.). The use of deterministic sources also reinforces the fact that the reference graphs are indeed assets that can be developed and curated to support a range of endeavors in addition to identity and entity resolution such as named entity recognition.

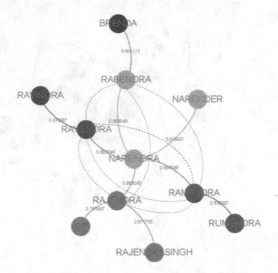

Fig. 2. This figure provides an example of Given Name Reference Graph (GNRG) annotation. The dashed blue lines represent the relationships that have been manually removed to ensure the four proper names "Rajendra", "Ravendra", "Rabendra", and "Ramendra" are discriminated between appropriately. Note how the community detection appropriately ascribes a different membership to each of these four names (Color figure online).

Next a non-overlapping community detection algorithm was deployed to partition the graph of names into classes (see Fig. 3. for an illustration of a subgraph of the FNRG). In this case the multilevel algorithm was used [19] due to its relative speed and performance.

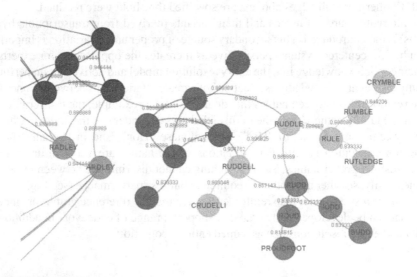

Fig. 3. This figure illustrates a subgraph of the Family Name Reference Graph (FNRG), including the membership classes derived from the community detection algorithm which are used in blocking.

Subsequent to the use of a community detection algorithm the classification derived from a proper name co-occurrence graph (for example, the name "John Edward Smith" would create a triad of "John– Edward; Edward – Smith; and John – Smith") derived from the target unresolved data, which is used to deterministically predict the origin of proper names using a coarse classification, is then used to allocate new hybrid classes.

Importantly the frequency of each name within the target unresolved dataset is retained as a node attribute to enable efficient assessment of the block sizes. The three levels of blocks in addition to the frequency of names from each block represented in the target unresolved dataset enables a degree of optimization, dependent on the domain specific context (e.g. the level of incompleteness and uncertainty in the data is significant), the business context (e.g. the cost of missing a match is high and real time assessment is not required so batch processing is preferable) and other factors such as hardware, software etc.

At this point the FNRG can be represented as a simple table of nodes with a membership integer, and the GNRG the more complex representation of a ragged array of integers, due to the nature of people having multiple given names.

2.2 Reference Graph Implementation

The entity resolution function with which the reference graphs were deployed within has four modules (see Fig. 4.).

The first module (see Fig. 4.) is Pre-Indexing which selects (e.g. Persons) and constrains (e.g. only those persons that have a Family name, Given name, and Date of Birth) the nodes to be compared for entity resolution. A range of parameters are used to configure this module.

The Equivalence Assessment module (see Fig. 4.) takes the set of entities from the previous module and creates sub-sets or blocks (Indexing), based on the algorithm selected as a parameter, to ensure the quadratic assessment of pairs is done in a scalable manner yet retaining as much accuracy as possible. Then Approximate String Matching (ASM), based on the algorithm (Jaro-Winkler or Cosine) selected as a parameter, is performed on each pair.

The Decision Management module (see Fig. 4.) takes the output of the previous module and a range of attributes from the target unresolved dataset (g) and makes a decision on whether the pair are a match or not. This decision is made on the basis of two concepts. Congruence – how similar the pair are in terms of the metrics available, and commonality – how unique the set of attributes are. These two concepts create the basis to not only decide whether the pair of entities are in fact the same real world entity in a probabilistic way but additionally how much certainty exists so optimized decision making can be made, given the domain context, of whether decisions can be made with a reduced set of attributes in a relatively inexpensive fashion, or whether uncertainty is high enough, given the domain context, to require an enriched set of attributes and a higher standard of proof. Decision making is conducted via a rule-based approach.

The Graph Contraction module (see Fig. 4.) manages the contraction of the graph using a range of methods including provenance to select what data to retain as the primary attributes. The meta-data derived from the ER model is retained.

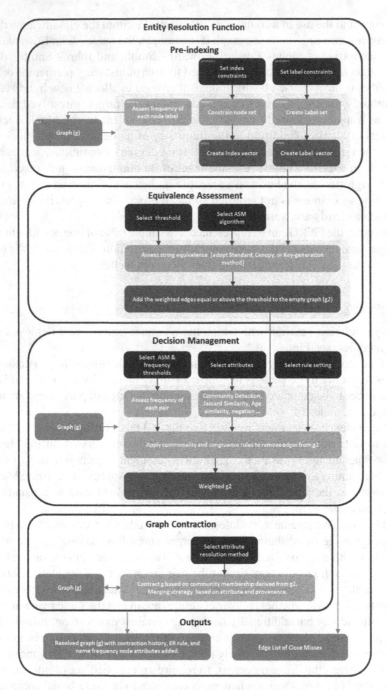

Fig. 4. This figure is a graphical representation of the design of the entity resolution function which was used to implement the experimental design. The indexing and decision management of the various strategies were conducted within the Equivalence Assessment and Decision Management modules respectively.

The reference graphs are used at both Equivalence Assessment (blocking) and Decision Management modules. Within the Equivalence Assessment module the FNRG community membership is used to create blocks or subsets of similar nodes to ensure the implementation is scalable. The granularity of blocks is dependent on the size and nature of the entity resolution task. The GNRG is not implemented within the Equivalence Assessment module at this point in time due to the added complexities of the overlapping nature of community membership.

However, both FNRG and GNRG are used in the Decision Management module. The FNRG is represented as a simple binary marker of whether the Family names of the pair of persons share the same class or not.

The GNRG is similarly implemented however rather than a binary approach each of the pairs community membership vectors (which can be of varying length dependent on how many given names they have) are compared and the number of matches is output.

3 Experiment

The utility of the reference graphs is measured on their ability to augment indexing and decision management. Performance measurement is based on the computational speed, scalability, and the number of true positive pairs identified through the application of one run of an optimally set entity resolution function. Four blocking strategies are compared. The reference graph strategy and three alternate blocking strategies have been deployed within the Equivalence Assessment module to serve as performance benchmarks. Each of the four blocking strategies is compared on the performance metrics under two decision management states – not using reference graph attributes and using reference graph attributes to support decision making. All other parameters are kept constant at near optimal levels for each dataset to replicate real-world conditions.

Importantly, the relative contextual performance of the overall ER model, which is comprised of multiple ER functions in a range of configurations, is briefly compared against a range of proprietary and non-proprietary entity resolution models that were benchmarked by Ferrante and Boyd [20]. This was done to provide comparative insight into how the reference graph algorithm contributes to the overall real world performance of the ER model.

3.1 Indexing Strategies

The "reference graph" strategy is implemented using the Family Name Reference Graph, and the classes derived from this graph.

The "Meta-Blocking Canopy" strategy [7–9] amalgamates blocks in a logical way to enhance accuracy with concomitant performance degradation. It is an effective optimizing strategy when the cost of inaccuracy is high, and computational expense can be sacrificed. The implementation here uses string length as the underpinning blocking strategy and then takes clusters of these blocks to form meta-blocks. The purpose for using the canopy strategy as a benchmark was to firstly to create a quasi-gold standard, and secondly to introduce the concept from a block-optimizing perspective to provide the explicit extension potential of using such strategies in tandem with reference graphs.

The second benchmark blocking strategy is a simple "Truncation" strategy. The implementation here is in two forms. For the small and two big datasets key-generation is derived from the truncation of each person entities family name at the first two letters (e.g. "DerekRobinson" > "RO") and first three letters (e.g. "DerekRobinson" > "ROB"), respectively. The reason for changing the implementation between small and large datasets is due to scalability and real world application.

The third strategy is the "Phonetic" strategy implemented using the Metaphone 3 (v2.5.4) algorithm [5] generating keys from each Person entities Family Name.

3.2   Decision Management States

The two decision management states implemented here consist of holding all other parameters constant and applying the use of the Given Name Reference Graph (GNRG) community membership attribute, or not.

3.3   Context

As discussed three real world target unresolved datasets are used to test the conditions previously outlined.

The first dataset is centered on Suspicious Transactions and is a small heterogeneous directed multiplex graph of approximately 40,000 nodes and 51,000 edges, comprised largely of manually annotated data, involving a vast range of relationships including familial, business, and financial transactions. The data contains high data incompleteness meaning there will be a high number of missing edges and nodes, and attributes thereof. Furthermore, fake and spoof nodes will also be present. Fake nodes are nodes that exist in the dataset that do not exist in the real world, and spoof nodes is where a real world entity is represented as multiple nodes in the data [2]. Both fake and spoof nodes will be derived from instances where real world entities intend to provide misinformation and where instances simply derive from human error. The cost of false positives and false negatives is high and therefore accuracy is paramount.

The second and third datasets are drawn from shareholding and directorship of NZ Companies. The Partial Companies graph is derived from the giant component from this data, and the Complete Companies graph is the complete graph. They are homogeneous multiplex weighted graphs of approximately 1.1 million nodes and 2.4 million edges, and 2.2 million nodes and 3.8 million edges respectively. The relationships consist of two types – shareholding and directorship. The data is drawn from the companies register and therefore is relatively complete; however there is a relatively low threshold of entity validation. Indeed it is highly probable that again fake and spoof nodes are present. The cost of false positives and false negatives is not as high, but scalability and run time become increasingly important considerations.

Here we are testing one entity resolution function to generate performance metrics from which to assess the utility of reference graphs. Of course within real world application the entity resolution function would be run, with alternate parameters, multiple times and as a collective form the implemented entity resolution model. In this experiment the performance metrics is couched within the context of the entity resolution model's performance and the context of each of the three target unresolved datasets.

It is also important to explicitly state that the entity resolution model was designed for batch process rather than real time, and the strength of the model is its extensibility and ability to be configured to incorporate a range of different methods at the indexing phase within the Equivalence Assessment module (e.g. Canopy, Phonetic, FNRG, Community Detection) and within the Decision Management module (e.g. hypocorisms, graph distance, reference graphs) for the purpose of entity resolving non-obvious latent entity pairs in the criminal context. Reference graphs are designed to ally the entity resolution of non-obvious pairs of entities and as such it is important to create the conditions where this is the goal. So, deduplication is conducted first to ensure all obvious entity resolutions are completed so as to leave only the non-obvious pairs.

In terms of the contextual performance of the ER model, of which the use of reference graphs is but one feature, both runtime and quality metrics (precision, recall, and f-measure) have been benchmarked against other ER models using Ferrante and Boyd's [20] comparison of software using synthetic data. Using the comparative performance results from Ferrante and Boyd's [20] study the ER model used here performs in the "slow" and "moderate" brackets for the small and large record sets respectively. However the quality metrics derived from the ER model used here (precision: 0.999, recall: 0.994, f-measure: 0.996) on the Suspicious Transactions Graph significantly outperform all benchmarked competitors (the top ranked software attained precision: 1.0, recall: 0.79, f-measure: 0.88), and using Ferrante and Boyd's classification would be classed as "very good". The limitations of this comparator are obvious but give a useful guide to performance for contextual purposes.

3.4 Performance Metrics

The following performance metrics have been used to assess the utility of the reference graphs within the experiment; computational expense; scalability and accuracy.

Computational expense consists of the measurement of each of the four blocking strategies across Equivalence Assessment (indexing) and Decision Management modules in both states (reference graph used in decision management or not) within the context of the entity resolution function and the overall entity resolution model.

Scalability is measured within the Equivalence Assessment module as the optimization of blocking remains a key research and applied problem. The primary metric is the number of computations required; however the number of blocks and the maximum number of entities within the blocks are important metrics to give a sense of block distribution, and how that distribution translates into computational performance.

Accuracy is measured simply by the number of pairs correctly resolved (true positives). The introduction of error through incorrectly resolving two entities within the context of the three data-sets generates a high cost and therefore the simple metric of counting the number of true positives is core. By-products of using a simple metric are that it was relatively simple to ensure each strategy was compared like for like, and the possibility of bias was reduced, and also the assessment of which strategy was most performant is made straight forward.

Diversity of the blocking approaches as a collective is another important concept to measure as in the real world the strategies are simply implemented as an alternate parameter setting from which the user can select and tune a specific entity resolution function. Not just one entity resolution function, but a number of entity resolution functions that together as a collective comprise the entity resolution model. Having said this however the partitions derived from the experiment indicate a nested structure, so those with a greater number of correctly resolved pairs are more diverse.

Table 1 outlines the scalability and accuracy of the reference graph and competing algorithms. The measurement of scalability was achieved by measuring the number of blocks generated, the maximum block size, the highest number ASM of computations conducted on a block, and total ASM computations. The measurement of accuracy was conducted by measuring the number of matches when the Given Name Reference Graph was not used in the Decision Management module, and the number of matches when the Given Name Reference Graph was used in the Decision Management module.

Table 2 illustrates the computational expense of the reference graph and competing algorithms across the three datasets for pre-processing, Equivalence Assessment (indexing) and Decision Management modules, the total run time for the Equivalence Assessment and Decision Management modules, and the average run time for each ER function.

3.5 Experimental Results

From Tables 1 and 2 the performance profiles of each strategy is evident. The Meta-Blocking Canopy strategy, as implemented, is not scalable as highlighted within scalability metrics and particularly the large number of total computations (80,103,833) but due to the near exhaustive equivalence assessment is most accurate on the Suspicious Transactions Graph with 317 (not using the GNRG) and 344 (using the GNRG) matches. In terms of its speed the Meta-Blocking Canopy strategy is slower, however the high accuracy of this approach means it remains a viable niche strategy on small graphs.

The Truncation strategy is scalable but performance will drop as the size of the dataset increases into the millions of nodes as either the number of computations increases sharply relative to the other algorithms or the truncation strategy is adjusted and accuracy suffers. Performance is good on the Suspicious Transactions Graph both in terms of expense and accuracy. This strategy is simple to implement, is fast on small graphs and relatively scalable.

The Phonetic (Metaphone 3) strategy, surprisingly, is consistently the poorest performer from an accuracy perspective, but the most scalable. The accuracy may be due to the very diverse set of names contained within the Suspicious Transactions graph, however the Companies graphs are less diverse and still the algorithm underperforms. Run times are slow on small graphs but relatively quicker on the larger graphs.

The reference graph strategy has a relatively expensive pre-processing time, however this one-off cost is offset under the context of the entity resolution model that will run multiple entity resolution functions together as a cluster to perform the resolution. Otherwise, the reference graph strategy consistently out-performed the other algorithms on both run time and accuracy. In terms of scalability the reference graph showed it is scalable, and due to being represented in a graph optimization is possible.

Table 1. Experimental results: Scalability and accuracy.

Suspicious Transactions Graph [40,000 nodes]

	Scalability				Accuracy	
	No. of Blocks	Max. Block size	Max. block computations	Total Computations	Matches (no GNRG)	Matches (GNRG)
Reference Graph	201	536	143,380	2,041,055	316	343
Meta-Blocking: Canopy	7	5,473	14,974,128	80,103,833	317	344
Truncation	307	553	152,628	1,173,819	314	337
Phonetic (Metaphone 3)	3,254	147	10,731	206,560	309	330

Partial Companies Graph [1,100,000 nodes]

	Scalability				Accuracy	
	No. of Blocks	Max. Block size	Max. block computations	Total Computations	Matches (no GNRG)	Matches (GNRG)
Reference Graph	2,194	8,230	33,862,335	602,590,934	4,351	5,954
Truncation	3,885	7,508	28,181,278	323,318,034	4,326	5,878
Phonetic (Metaphone 3)	4,704	4,914	12,071,241	204,704,633	4,314	5,842

Complete Companies Graph [2,200,000 nodes]

	Scalability				Accuracy	
	No. of Blocks	Max. Block size	Max. block computations	Total Computations	Matches (no GNRG)	Matches (GNRG)
Reference Graph	3,322	10,667	56,887,111	1,181,286,210	29,555	34,730
Truncation	4,847	14,380	103,385,010	1,402,409,226	28,626	33,343
Phonetic (Metaphone 3)	5,118	7,599	28,868,601	804,107,679	28,233	33,008

Table 2. Experimental results: Computational expense (run time in seconds).

Suspicious Transactions Graph [40,000 nodes]: Approx. ER Model run time 630 seconds

No use of the Reference Graph in Decision Management

	Pre-processing (sec)	Equiv. Assessment (sec)	Decision Mgmt (sec)	Total (sec)	Average ER Function (sec)
Reference Graph	150.25	5.27	1.34	6.61	7
Meta-Blocking: Canopy	0	65.35	1.84	67.19	67
Truncation	0.74	9.89	1.65	11.55	13
Phonetic (Metaphone 3)	4.61	58.26	1.60	59.93	66

Use of the Reference Graph in Decision Management

	Pre-processing (sec)	Equiv. Assessment (sec)	Decision Mgmt (sec)	Total (sec)	Average ER Function (sec)
Reference Graph	150.25	5.02	1.68	6.70	8
Meta-Blocking: Canopy	0	64.80	1.63	66.42	67
Truncation	0.74	11.03	1.66	12.69	13
Phonetic (Metaphone 3)	4.61	64.56	1.71	66.27	67

Partial Companies Graph [1,200,000 nodes] : Approx. ER Model run time 4 hours

No use of the Reference Graph in Decision Management

	Pre-processing (sec)	Equiv. Assessment (sec)	Decision Mgmt (sec)	Total (sec)	Average ER Function (sec)
Reference Graph	524	3,286	59	3,501	3,726
Truncation	11	5,032	61	5,110	5,337
Phonetic (Metaphone 3)	28	4,124	59	4,183	4,375

Use of the Reference Graph in Decision Management

	Pre-processing (sec)	Equiv. Assessment (sec)	Decision Mgmt (sec)	Total (sec)	Average ER Function (sec)
Reference Graph	524	3,301	63	3,339	3,562
Truncation	11	4,983	67	5,010	5,277
Phonetic (Metaphone 3)	28	4,630	59	4,689	4,903

Complete Companies Graph [2,200,000 nodes] : Approx. ER Model run time 5.1 hours

No use of the Reference Graph in Decision Management

	Pre-processing (sec)	Equiv. Assessment (sec)	Decision Mgmt (sec)	Total (sec)	Average ER Function (sec)
Reference Graph	1,486	7,641	255	7,897	8,221
Truncation	16	9,397	262	9,659	9,994
Phonetic (Metaphone 3)	51	11,321	254	11,575	11,783

Use of the Reference Graph in Decision Management

	Pre-processing (sec)	Equiv. Assessment (sec)	Decision Mgmt (sec)	Total (sec)	Average ER Function (sec)
Reference Graph	1,486	7,630	251	7,881	8,101
Truncation	16	9,407	259	9,666	9,865
Phonetic (Metaphone 3)	51	11,352	262	11,614	11,710

Perhaps the most significant finding was the clear advantage of using the Given Name Reference Graph to assist making decisions, especially considering there is no material computational expense involved.

4   Discussion

The experimental results clearly mark the applied utility of the reference graph strategy, and excitingly, the demonstrated applied utility is buttressed by a number of features that extends the real-world value of this strategy.

The experimental results clearly indicate the relative scalability, expense, and accuracy of the reference graph as a blocking strategy, across all graph types examined showing encouraging performance and generalized applicability. Furthermore, the reference graph has a number of features that extends this value under real-world conditions. From a human centered computing perspective the reference graph can be improved over time by the curation of human experts annotating the relationships between proper names, crucially including ensuring counter-factual relationships between proper names do not exist (e.g."Rabendra" ! = "Ravendra"), and validate the performance of community detection.

The flexibility of the coarseness of partitioning, or indeed potential for over-lapping classes, is another feature that enable meta-blocking like capabilities creating the opportunity to tune the reference graph dependent on the contextual requirements demanded by each individual set of instances.

As an adjunct to decision management the case for the use of reference graphs is compelling. Performance was significantly enhanced with little to no material increase in expense. From a criminal network perspective, as alluded to earlier, performance enhancements targeting the non-obvious pairs is the focus and a very complex and challenging problem. The results derived from the experiment are very encouraging from this perspective, both in terms of dealing with the higher uncertainty of the Suspicious Transactions Graph, and from a scalability perspective with the Companies Graphs. Of course these are only indicative findings and further comparison against a variety of "state of the art" algorithms using a diverse range of criminal datasets is required to further validate the utility of reference graphs.

From a real world perspective the application of the ER model using reference graphs within the criminal domain has significantly improved downstream models, designed to detect, measure and prioritize risk, that consume the output from the ER model. It was expected that the downstream benefit would be more amplified in anomaly detection approaches, however there has proven to be a significantly improved performance across the more generic models as well. Models such as shortest-path based models that identify subgraphs of clusters of entities involved in suspicious transactions that are linked to entities that generate illicit income (e.g. methamphetamine traffickers), and graph propagation models used across a range of criminal sub-domains. This provides initial tentative support to the earlier assertion that the latent non-obvious pairs are of most importance, and indeed this small set of latent actors may well be essential to the ongoing criminal structural fabric of criminal networks.

5 Extensions

A number of areas have been identified to extend the current implementation of reference graphs.

The use of overlapping community detection approaches will undoubtedly increase the accuracy of the reference graphs, however there is certainly a trade-off of making the approach more complex and potentially significantly more expensive even if data representations like hypergraphs are used.

The reference graphs are used as a binary attribute to drive blocking and support decision management. However, there is the simple extension of deploying the attribute as a graph-distance based metric that creates the opportunity for optimized blocking via a meta-blocking implementation and at the decision management phase creates the opportunity to use a more nuanced and sophisticated approach to support making decisions.

6 Conclusion

The use of reference graphs to bolster performance in entity resolution, at both indexing and decision management stages, has been clearly demonstrated within this paper, with both experimental results and the outlining of additional real-world benefits. This coupled to the reference graphs wide applicability, simple implementation, and numerous areas for extensions points to an entity resolution strategy that has great potential for generating real-world value.

Specifically within the criminal domain the use of reference graphs within entity resolution has been demonstrated to be both performant from an accuracy perspective, which is critical when targeting non-obvious instances, and also performant from a scalability perspective. This unlocks the ability to federate data between a criminal network hub and multiple large heterogeneous datasets (the spokes), in addition to providing quality accurate resolution with data characterised by incompleteness, high uncertainty, and the presence of fake and spoof nodes.

References

1. Benjelloun, O., Garcia-Molina, H., Menestrina, D., Su, Q., Whang, S.E., Widom, J.: Swoosh: a generic approach to entity resolution. VLDB J. **18**(1), 255–276 (2009)
2. Maeno, Y.: Node discovery problem for a social network. Connections **29**, 62–76 (2009)
3. Odell, M., Russell, R.: The Soundex Coding System. US Patents 1261167 (1918)
4. Philips, L.: The double metaphone search algorithm. C/C ++ Users J. **18**(6), 38–43 (2000)
5. Philips, L.: Metaphone 3 version 2.5.4 (2015)
6. de Vries, T., Ke, H., Chawla, S., Christen, P.: Robust record linkage blocking using suffix arrays. In: Proceedings of the 18th ACM Conference on Information and Knowledge Management, pp. 305–314. ACM (2009)
7. Hernández, M.A., Stolfo, S.J.: The merge/purge problem for large databases. In: Proceedings of the ACM SIGMOD International Conference on Management of Data 1995, pp. 127–138. ACM, New York (1995)

8. Hernández, M.A., Stolfo, S.J.: Real-world data is dirty: data cleansing and the merge/purge problem. Data Min. Knowl. Discov. **2**(1), 9–37 (1998)
9. McCallum, A., Nigam, K., Ungar, L.H.: Efficient clustering of high-dimensional data sets with application to reference matching. In: Proceedings of the Sixth ACM SIGKDD International Conference on Knowledge Discovery and Data Mining, pp. 169–178. ACM (2000)
10. Taylor, J.: Decision Management Systems: A Practical Guide to Using Business Rules and Predictive Analytics. Pearson Education, Boston (2012)
11. Bhattacharya, I., Getoor, L.: Collective entity resolution in relational data. ACM Trans. Knowl. Discov. Data **1**(1), 1–36 (2007)
12. Bhattacharya, I., Getoor, L.: Entity Resolution in Graphs. In: Cook, D.J., Holder, L.B. (eds.) Mining Graph Data, pp. 311–344. Wiley, Hoboken (2006)
13. Köpcke, H., Rahm, E.: Frameworks for entity matching: a comparison. Data Knowl. Eng. **69**(2), 197–210 (2010)
14. Randall, S.M., Boyd, J.H., Ferrante, A., Bauer, J.K., Semmens, J.B.: Use of graph theory measures to identify errors in record linkage. Comput. Methods Programs Biomed. **115**(2), 55–63 (2014)
15. Zhou, Y., Talburt, J.R.: Strategies for large-scale entity resolution based on inverted index data partitioning. In: Yeoh, W., Talburt, J.R., Zhou, Y. (eds.) Information Quality and Governance for Business Intelligence, pp. 329–351. IGI Global, Hershey (2013)
16. Michalowski, M., Thakkar, S., Knoblock, C.A.: Exploiting secondary sources for unsupervised record linkage. In: Proceedings of the 30th VLDB Conference, Toronto, Canada (2004)
17. Papadakis, G., Koutrika, G., Palpanas, T., Nejdl, W.: Meta-blocking: taking entity resolution to the next level. IEEE Trans. Knowl. Data Eng. **26**(8), 1946–1960 (2014)
18. Winkler, W.E.: String comparator metrics and enhanced decision rules in the fellegi-sunter model of record linkage. In: Proceedings of the Section on Survey Research Methods, American Statistical Association, pp. 354–359 (1990)
19. Blondel, V.D., Guillaume, J.-L., Lambiotte, R., Lefebvre, E.: Fast unfolding of communities in large networks. J. Stat. Mech. Theory Exper. **2008**(10), P10008 (2008)
20. Ferrante, A., Boyd, J.: A transparent and transportable methodology for evaluating data linkage software. J. Biomed. Inform. **45**(1), 165–172 (2012)

Heterogeneous Information Networks Bi-clustering with Similarity Regularization

Xianchao Zhang, Haixin Li, Wenxin Liang$^{(\boxtimes)}$, Linlin Zong, and Xinyue Liu

Dalian University of Technology, Dalian, China
{xczhang,wxliang,xyliu}@dlut.edu.cn,
{lihaixin,linlinzong}@mail.dlut.edu.cn

Abstract. Clustering analysis of multi-typed objects in heterogeneous information network (HINs) is an important and challenging problem. Nonnegative Matrix Tri-Factorization (NMTF) is a popular bi-clustering algorithm on document data and relational data. However, few algorithms utilize this method for clustering in HINs. In this paper, we propose a novel bi-clustering algorithm, BMFClus, for HIN based on NMTF. BMFClus not only simultaneously generates clusters for two types of objects but also takes rich heterogeneous information into account by using a similarity regularization. Experiments on both synthetic and real-world datasets demonstrate that BMFClus outperforms the state-of-the-art methods.

Keywords: Heterogeneous information network · Nonnegative Matrix Tri-Factorization · Clustering

1 Introduction

Many real world data can be modeled using heterogeneous information networks (HINs) which consists of multiple types of objects. For example, a heterogeneous bibliographic network in Fig. 1(a) contains multi-typed objects including *authors*, *venues* (conferences or journals) and *terms*. A heterogeneous servers network in Fig. 1(d) contains *switches, email servers, database servers* and *web servers*. Clustering in HINs has attracted increasing attention in recent years. For instance, [1] finds that clustering analysis in heterogeneous bibliographic network helps generate more effective and comprehensive ranking result for *authors* and *venues*. In a heterogeneous servers network, if an attack script runs on some compromised web servers and the script reads data from database servers and sends out spam emails through the email servers, we call these servers compose an "attack sub-network". [2] proposes a framework to find such "attack sub-networks" with the help of clustering in HINs.

However, most existing studies on HIN clustering have some limitations. Many studies [3–5] deal with HINs as homogeneous networks, i.e., a network

---

This work was supported by National Science Foundation of China (No. 61272374,61300190) and 863 Project (No. 2015AA015463).

© Springer International Publishing Switzerland 2016
M. Chau et al. (Eds.): PAISI 2016, LNCS 9650, pp. 19–30, 2016.
DOI: 10.1007/978-3-319-31863-9_2

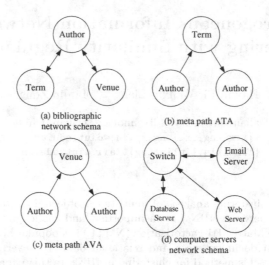

Fig. 1. Example of network schemas and meta paths

consisting of single type of objects. All the types (author type, venue type or term type) are treated in the same way in these algorithms. Therefore, these algorithms fail to use rich heterogeneous information during the clustering process. Some studies, RankClus [1] and GenClus [6] for example, distinguish different types in HINs. They assign a certain type in HIN as target type and other types as attribute types. Then, they focus on clustering target type and only generate clusters for that type of objects. HIN clustering aims at finding K partitions for multi-typed objects, so that objects in the same partition should be more similar (have more connections) to each other than to those in other partitions. For example, in a heterogeneous bibliographic network, we can tell which venues belong to "Information Security" and which venues belong to "Information Retrieval" by using clustering analysis in venues. While, if we are interested in which authors are authorities in "Information Security" and which authors are authorities in "Information Retrieval", we have to cluster the authors. Therefore, a better way to analysis heterogeneous bibliographic networks is to generate clusters for venues and authors simultaneously. Unlike RankClus [1] and GenClus [6], we try to find partitions for more than one type of objects. In this case, bi-clustering is a feasible technique because it generates clusters for two types of objects. In recent years, bi-clustering based on Nonnegative Matrix Tri-Factorization (NMTF) [7,8] attracts increasing attention because of their mathematical elegance and encouraging empirical results on document data and relational data. Nevertheless, few algorithms utilize NMTF for the HIN clustering. The main challenge of applying NMTF to HIN is how to incorporate rich heterogeneous information into the clustering process.

To address the problem, in this paper we propose a novel bi-clustering algorithm, BMFClus (HIN Bi-Clustering based on Matrix tri-Factorization), which simultaneously generates clusters for two types of objects, and takes advantage of rich heterogeneous information during the clustering process. To achieve this goal, the NMTF is adopted as a basic bi-clustering method of BMFClus to cluster two types of objects in HIN. Furthermore, a similarity regularization term is introduced to the objective function of NMTF. The similarity regularization term enables BMFClus to utilize rich heterogeneous information in HINs, which leads to an improvement over the basic NMTF method. Our contributions are summarized as follows:

1. We propose a bi-clustering algorithm for HINs based on NMTF.
2. We incorporate rich heterogeneous information into the bi-clustering process by a similarity regularization term.
3. Experiments on both synthetic and real-world datasets demonstrate the effectiveness of the proposed algorithm in comparison with the state-of-the-art algorithms.

The rest of this paper is organized as follows: Sect. 2 introduces the problem statement and related work. Section 3 describes the details of the proposed algorithm. Section 4 reports the performance of the proposed algorithm comparing with the state-of-the-art algorithms. Finally, we concludes the paper and outlines the future work.

2   Problem Statement and Related Work

A graph $G = (V, E)$, where $V = \bigcup_{i=1}^{t} X_i$, and $X_1 = \{x_{11}, ..., x_{1n_1}\}, ..., X_t = \{x_{t1}, ..., x_{tn_t}\}$ denote the t different types of nodes. E is the set of links between any two data objects in V. If $t = 1$, G is a **homogeneous information network**. If $t > 1$, G is a **heterogeneous information network**.

As described in [9], a graph $S = (T, R)$ is called **network schema**, if S is an undirected connected graph defined over object types T, with edges as relations from R. A network schema provides a meta structure description of a heterogeneous information network.

For example, Fig. 1 (a) is a network schema of a HIN, specifically, a heterogeneous bibliographic network. It contains three types of objects including *authors*, *venues* and *terms*. For this HIN, $T = \{Author(A), Venue(V), Term(T)\}$, $R = \{A - V, V - A, A - T, T - A\}, t = 3$, X_1 denotes the objects of author type, X_2 denotes the objects of venue type, X_3 denotes the objects of term type. We define the **meta path** as a path in network schema which connects two types, following the definition of meta path in [9]. In Fig. 1, (b) and (c) are two meta paths selected from (a). Meta path (b) composed by relations $A - T$ and $T - A$, and is denoted as ATA. ATA encodes the semantic that whether two authors are interested in the same term, e.g. both two authors like "kmeans". Meta path (c), i.e. AVA, composed by relations $A - V$ and $V - A$, denotes the semantic that whether two authors are interested in the same venue, e.g. two co-authors publish a paper in "SIGKDD".

Given a meta path, we can use **PathCount** to measure the similarity between a pair of objects [9]. The PathCount of x_{1i}, x_{1j} is the number of path from x_{1i}

to x_{1j} following a certain meta path. For instance, in Fig. 1, two authors can be connected via "author-term-author" (ATA) path if they use a same term in their papers, and $PathCount(x_{1i}, x_{1j})$ under path ATA is the number of common terms used by author x_{1i} and x_{1j}. Meta path "author-venue-author" (AVA) denotes a relation between authors via venues (i.e., publishing in the same venues), and $PathCount(x_{1i}, x_{1j})$ under path AVA is the number of common venues attended by author x_{1i} and x_{1j}. Given a meta path, the higher value of $PathCount(x_{1i}, x_{1j})$, x_{1i} and x_{1j} are considered to be more similar. Since meta path encodes the relationship between different types, it captures rich heterogeneous information of a HIN [9–11].

Several approaches have been proposed to find K partitions for the multi-typed objects in a HIN. SpectralBiclustering [12] is proposed to bi-clustering two types of objects using spectral clustering. RankClus [1] combines ranking and clustering techniques to analysis two types of objects in HIN. PathSelClus [11] utilizes rich heterogeneous information encoded by meta path. While, PathSel-Clus only generates cluster for a single type in HIN. Our work is different from theirs, as we focus on simultaneously generating clusters for two types of objects. In addition, we also propose how to incorporate rich heterogeneous information into the NMTF clustering process.

3   Proposed Algorithm

3.1   NMTF

We give a brief review of Nonnegative Matrix Tri-Factorization (NMTF) [8] which is an effective bi-clustering method. Given a data matrix $M \in \mathbb{R}_+^{m \times n}$, the objective function of the NMTF is

$$\min_{F \geq 0, G \geq 0, S \geq 0} \left\| M - FSG^T \right\|_F^2, \tag{1}$$

where $\|\cdot\|_F$ denotes the matrix Frobenius norm, $F \in \mathbb{R}_+^{m \times k}$, $S \in \mathbb{R}_+^{k \times k}$ and $G \in \mathbb{R}_+^{n \times k}$. S provides additional degrees of freedom such that the low-rank matrix representation remains accurate, while F gives m row cluster assignment vectors and G gives n column cluster assignment vectors. Equation (1) can be computed using the following update rules [8].

$$G_{jk} \leftarrow G_{jk} \sqrt{\frac{(M^T FS)_{jk}}{(GG^T M^T FS)_{jk}}}, \tag{2}$$

$$F_{ik} \leftarrow F_{ik} \sqrt{\frac{(MGS^T)_{ik}}{(FF^T MGS^T)_{ik}}}, \tag{3}$$

$$S_{ik} \leftarrow S_{ik} \sqrt{\frac{(F^T MG)_{ik}}{(F^T FSG^T G)_{ik}}}. \tag{4}$$

Although the objective function in Eq. (1) is not convex in all variables together, it is proved that the above update rules will find a local minimum of Eq. (1). Using NMTF for data clustering has following merits:

1. We can obtain the clusters of rows and columns simultaneously of a data matrix. Actually, it is also proved that NMTF is equivalent to do kernel K-means clustering on both columns and rows [8].
2. NMTF conducts a knowledge transformation between the row feature space and the column feature space [13]. It means that the quality of the row clustering and the column clustering are mutually enhanced during the update iteration.

The performance of NMTF in document data or relation data has been well studied. However, to the best of our knowledge, we are the first to apply NMTF to HIN clustering.

3.2 BMFClus

Although NMTF can be used to generate clusters for two types on a HIN, it fails to take advantage of rich heterogeneous information captured by meta path. Therefore, we propose BMFClus which not only inherits advantages of NMTF, but also takes into account rich heterogeneous information of HIN.

First, we use NMTF to model a HIN. Two types are selected from T. Then, a nonnegative edges weight matrix M is constructed, where $M_{i,j}$ is the number of the links between two nodes. For example, in Fig. 1, we choose *author* type and *venue* type. The $M_{i,j}$ denotes how many papers of author i published by venue j. If the topic of venue j is "data mining", and the value of $M_{i,j}$ is high, then author i has a high probability to be labeled as "data mining", and vice versa. According to Eq. (1), F gives the cluster assignment vectors of authors and G gives the cluster assignment vectors of venues. If $k = 3$ and $F_{i,*} = [0.9, 0.1, 0]$, then author i will be assigned to cluster 1. If $G_{*,j} = [0.05, 0.95, 0.05]^T$, then venue j will be assigned to cluster 2.

Next, we describe how to incorporate rich heterogeneous information into NMTF. As mentioned before, meta paths encode rich heterogeneous information of HIN. Given a meta path, a nonnegative similarity matrix is constructed using PathCount. For example, in Fig. 1, if we use meta path $p = ATA$ (author-term-author), a nonnegative similarity matrix $W^{(F)}$ between each authors can be constructed using $W_{i,j}^{(F)} = PathCount(x_{1i}, x_{1j})$. And $W_{i,j}^{(F)}$ is the number of path from object x_{1i} to object x_{1j} following p.

Our goal is to encourage two objects (x_{1i} and x_{1j}) who have a high similarity ($W_{i,j}^{(F)}$) to have similar cluster assignment vectors ($F_i \approx F_j$). To achieve this goal, we introduce the following similarity regularization term.

$$O_1 = \frac{1}{2} \sum_{i,j} \|F_i - F_j\|_2^2 W_{i,j}^{(F)}. \tag{5}$$

The regularization term Eq. (5) is a cost function. It is obvious that if x_{1i} and x_{1j} have a high similarity value with respect to $W_{i,j}^{(F)}$, we should make $\|F_i - F_j\|_2^2$ small to reduce the punishment of Eq. (5). Minimizing O_1 will smooth the cluster distributions between a object and its similar objects. Define diagonal matrix $D_{i,i}^{(F)} = \sum_j W_{i,j}^{(F)}$. Then, we construct the consistent Laplacian matrix $L_F = D^{(F)} - W^{(F)}$. Now, we rewrite the regularization term into trace form:

$$
\begin{aligned}
O_1 &= \frac{1}{2} \sum_{i,j} \|F_i - F_j\|_2^2 W_{i,j}^{(F)} \\
&= \sum_i F_i D_{i,j}^{(F)} F_i^T - \sum_{i,j} F_i W_{i,j}^{(F)} F_j^T \\
&= Tr(F^T L_F F).
\end{aligned}
\tag{6}
$$

Similar with the construction of O_1, we construct another objective function for venue type:

$$
O_2 = Tr(G^T L_G G).
\tag{7}
$$

Now, we define our **BMFClus** by adding the regularization terms Eqs. (6) and (7) to Eq. (1):

$$
\min_{F \geq 0, G \geq 0, S \geq 0} \left\| M - FSG^T \right\|_F^2 + \lambda (Tr(F^T L_F F) + Tr(G^T L_G G)),
\tag{8}
$$

where the first term represents the reconstruction error for nonnegative edges weight matrix M, and the second term represents the similarity regularization. λ is a trade-off parameter. This parameter is not much sensitive and we set it to be 0.1 in our experiments.

Algorithm 1. *BMFClus*

Require: M, $W^{(F)}$ and $W^{(G)}$ constructed using HIN,
         trade-off parameters λ
         number of clusters K
Ensure: the cluster assignment vectors F and G
 1: Normalize M, $W^{(F)}$ and $W^{(G)}$;
 2: Initialize F, S and G ;
 3: **repeat**
 4:     Fixing other factors, update S by Eq. (11);
 5:     Fixing other factors, update F by Eq. (10);
 6:     Fixing other factors, update G by Eq. (9);
 7: **until** Eq. (8) converges.

BMFClus can be computed using the following update rules [14]:

$$
G_{jk} \leftarrow G_{jk} \sqrt{\frac{[\lambda L_G^- G + A^+ + GB^-]_{jk}}{[\lambda L_G^+ G + A^- + GB^+]_{jk}}},
\tag{9}
$$

$$F_{ik} \leftarrow F_{ik} \sqrt{\frac{[\lambda L_F^- F + P^+ + FQ^-]_{ik}}{[\lambda L_F^+ F + P^- + FQ^+]_{ik}}}, \tag{10}$$

$$S = (F^T F)^{-1} F^T M G (G^T G)^{-1}, \tag{11}$$

where $L_G = L_G^+ - L_G^-$, $A = M^T F S = A^+ - A^-$, $B = S^T F^T F S = B^+ - B^-$, $P = MGS^T$, $Q = SG^T GS^T$, $A_{ij}^+ = (|A_{ij}| + A_{ij})/2$, $A_{ij}^- = (|A_{ij}| - A_{ij})/2$ [15]. The detailed process of the BMFClus is summarized in Algorithm 1.

4   Experiments

In this section, we conduct a series of experiments to show the effectiveness of BMFClus on both synthetic and real datasets.

4.1   Dataset

We give a brief description of the datasets used in our experiments as follows:

1. **SynData**: We generate a synthetic HIN following the properties of real word HIN. A HIN is composed with several bipartite networks. We apply the method described in [1] to generate 3 synthetic bipartite networks and construct a synthetic HIN, SynData. SynData contains 3 clusters and 3 types, denoted as A, B and C. The number of objects: $N_a = \{1000, 1200, 1300\}$ for type A, $N_b = \{3000, 3200, 3500\}$ for type B, $N_c = \{1000, 1200, 1300\}$ for type C.
2. **DBLP4** [16] is a real-world HIN extracted from DBLP bibliography dataset in four research areas: database (DB), data mining (DM), information retrieval (IR) and artificial intelligence (AI). DBLP4 contains 3 types of objects: 4236 authors, 20 venues and 11771 unique terms. Each author object links with several venues and terms. The link weight of author-venue pair is the number of papers the author publishes in the venue. The author-term sub-network contains all the terms appeared in the abstract of papers of each author with stopwords removed. All the venue objects and author objects are labeled.
3. **Flickr** [9,17]: Flickr is a HIN contains three types of objects: image, user and tag. Each image object links with several tags and one user. Image objects are labeled.

The statistics of the datasets are summarized in Table 1. The network schemas of the datasets are shown in Fig. 2.

Table 1. Statistics of the datasets

dataset	# object	# type	# link	# cluster
SynData	16700	3	135000	3
DBLP4	16027	3	735710	4
Flickr	4076	3	14396	8

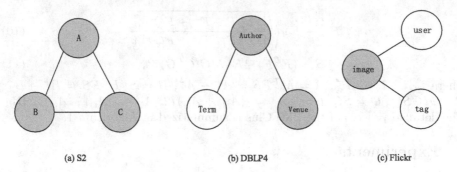

Fig. 2. Network schemas of the datasets. (a):SynData; (b):DBLP4; (c): Flickr. The labeled object types are in grey.

4.2 Baselines

We compare the proposed method with the following state-of-the-art algorithms:

- **SpectralBiclustering (SBC)** [12] is a well known spectral clustering based bi-clustering algorithm. Give a $m \times n$ matrix, SBC generates clusters for m rows and n columns.
- **GreedyCoClustering** [9] is an information-theoretic greedy bi-clustering algorithm. It use a greedy KL-divergence based bi-clustering method to cluster a $m \times n$ matrix.
- **NMTF** [8] is the basic form of the proposed method.
- **RankClus** [1] is a rank-based algorithm which integrates ranking and clustering together for a heterogeneous bibliographic network. RankClus treats a $m \times n$ matrix as a bipartite graph. RankClus generates clusters for m rows, and applies ranking for n columns. Then, it generates a better cluster structure for m rows based on the ranking distribution on n columns. After several this iteration, quality of clustering and ranking are mutually enhanced.

4.3 Evaluation Metrics

The clustering performance is evaluated by comparing the ground truth labels with the predicted labels. Two popular metrics, i.e., *accuracy* (ACC) and *normalized mutual information* (NMI), are used to measure the clustering performance [1,18,19].

Given an object v_i of a certain type T_a $(1 \leq a \leq t)$, let c_i and r_i be the predicted label and the ground truth label of v_i, respectively. The ACC of type T_a is defined as follows:

$$ACC = \frac{\sum_{i=1}^{n_a} \delta\left(c_i, map\left(r_i\right)\right)}{n_a}. \qquad (12)$$

where $\delta(x,y)$ equals one if $x = y$ and equals zero otherwise. $map(r_i)$ is the permutation mapping function that maps each cluster label r_i to the equivalent label from the ground truth labels. Kuhn-Munkres algorithm [20] is used for finding the best mapping.

Given the clustering result of type T_a, let $n(i,j), i,j = 1,2,...,K$, denote the number of objects that predicted as label i and labeled as j in the ground truth. From $n(i,j)$, we define joint distribution $p(i,j) = \frac{n(i,j)}{n_a}$, row distribution $p_1(j) = \sum_{i=1}^{K} p(i,j)$ and column distribution $p_2(i) = \sum_{j=1}^{K} p(i,j)$. The NMI of type T_a is defined as follows:

$$NMI = \frac{\sum_{i=1}^{K}\sum_{j=1}^{K} p(i,j) \log(\frac{p(i,j)}{p_1(j)p_2(i)})}{\sqrt{\sum_{j=1}^{K} p_1(j) \log p_1(j) \sum_{i=1}^{K} p_2(i) \log p_2(i)}}. \tag{13}$$

NMI dose not require the mapping function between the predicted labels and ground truth labels.

Both metrics are in the range from 0 to 1 and a higher value indicates a better clustering performance in terms of the ground truth labels.

4.4 Settings

For DBLP4 dataset, *author* type and *venue* type are selected as the two types of objects we want to cluster. And the nonnegative edges weight matrix M is constructed by the link count between two objects. L_F and L_G is constructed using meta path ATA (author-term-author) and VAV (venue-author-venue), respectively. Meta path ATA means two authors share more common terms are more similar and meta path VAV means two venues share more common authors are more similar. For Flickr dataset, we choose *image* type and *tag* type as the types of objects we want to cluster. M is constructed by the links between image objects and tag objects. Since only image objects are labeled, the clustering performance are evaluated on image type. L_F and L_G is constructed using meta path IUI (image-user-image) and TIT (tag-image-tag), respectively. Meta path IUI means two images provided by same user are similar, and meta path TIT means two tags share more common images are more similar. For SynData dataset, we choose A type and B type as the clustering target. M is constructed by the links between A and B. L_F and L_G is constructed using meta path ACA and BCB, respectively. M also serves as the input data for SBC [12], GreedyCoClustering [9], NMTF [8] and RankClus [1]. Each result is the average of 10 runs.

4.5 Results

The results are shown in Table 2 and the best results are highlighted in bold-face. On the DBLP4 and SynData dataset, as this dataset is nicely structured, all methods achieve outstanding performance. NMTF outperforms the other

Table 2. Cluster performance of different methods.

(a) ACC

dataset	DBLP4		Flickr	SynData	
type	author	venue	image	A	B
SBC [12]	0.8780	0.8500	**0.4669**	0.8803	0.7185
GreedyCoClustering [9]	0.7652	0.7668	0.3597	0.8207	0.6621
NMTF [8]	0.9308	0.9500	0.4205	0.9222	0.7850
RankClus [1]	0.6822	0.8500	0.4161	0.9316	0.7731
BMFClus	**0.9327**	**1.0000**	0.4317	**0.9438**	**0.8153**

(b) NMI

dataset	DBLP4		Flickr	SynData	
type	author	venue	image	A	B
SBC [12]	0.7279	0.8388	0.3997	0.7024	0.3518
GreedyCoClustering [9]	0.5812	0.6628	0.2419	0.6557	0.3021
NMTF [8]	0.7834	0.9058	0.4131	0.7328	0.3924
RankClus [1]	0.5931	0.8338	0.3934	0.7191	0.3889
BMFClus	**0.7901**	**1.0000**	**0.4242**	**0.7618**	**0.4137**

baselines on DBLP4 and SynData dataset. As expected, due to the rich heterogeneous information captured by similarity regularization term, BMFClus performs much better than NMTF on author type and venue type with respect to NMI and ACC. On the Flickr dataset, we observe that SBC [12] achieves the best accuracy on image type. While, BMFClus outperforms SBC with respect to NMI. BMFClus achieves the best results on DBLP4 and SynData dataset, and the second best results on Flickr dataset. Overall we conclude that BMFClus outperforms the base line methods.

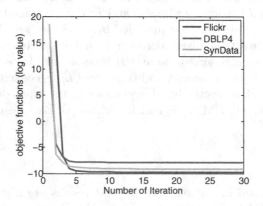

Fig. 3. Convergence of BMFClus

4.6   Algorithm Convergence

The update rules for minimizing the objective functions of BMFClus are essentially iterative We investigate the convergence of BMFClus. Figure 3 shows the convergence curve of the objective functions (in log values) on three datasets.

It is easy to see that the objective values of BMFClus falling fast at the first several iterations on each datasets.

5   Conclusion and Future Work

In this paper, we propose a bi-clustering algorithm (BMFClus) for HINs based on NMTF. Specifically, given a HIN, BMFClus simultaneously generates clusters for two types of objects. Besides, BMFClus takes rich heterogeneous information into account by using a similarity regularization. Experiments on both synthetic and real-world datasets demonstrate that BMFClus outperforms the state-of-the-art methods. For the future work, we will investigate how to extend BMFClus to arbitrary multi-typed heterogeneous information networks.

References

1. Sun, Y., Han, J., Zhao, P., Yin, Z., Cheng, H., Wu, T.: Rankclus: integrating clustering with ranking for heterogeneous information network analysis. In: Proceedings of the 12th International Conference on Extending Database Technology: Advances in Database Technology, pp. 565–576. ACM (2009)
2. Gupta, M., Gao, J., Yan, X., Cam, H., Han, J.: Top-k interesting subgraph discovery in information networks. In: 2014 IEEE 30th International Conference on Data Engineering (ICDE), pp. 820–831. IEEE (2014)
3. Wang, N., Parthasarathy, S., Tan, K.-L., Tung, A.K.: Csv: visualizing and mining cohesive subgraphs. In: Proceedings of the 2008 ACM SIGMOD International Conference on Management of Data, pp. 445–458. ACM (2008)
4. White, S., Smyth, P.: A spectral clustering approach to finding communities in graph. In: SDM, vol. 5, pp. 76–84. SIAM (2005)
5. Liu, X., Yu, S., Janssens, F., Glänzel, W., Moreau, Y., De Moor, B.: Weighted hybrid clustering by combining text mining and bibliometrics on a large-scale journal database. J. Am. Soc. Inform. Sci. Technol. 61(6), 1105–1119 (2010)
6. Sun, Y., Aggarwal, C.C., Han, J.: Relation strength-aware clustering of heterogeneous information networks with incomplete attributes. Proc. VLDB Endowment 5(5), 394–405 (2012)
7. Pei, Y., Chakraborty, N., Sycara, K.: onnegative matrix tri-factorization with graph regularization for community detection in social networks. In: Proceedings of the 24th International Conference on Artificial Intelligence, IJCAI 2015, pp. 2083–2089. AAAI Press (2015)
8. Ding, C., Li, T., Peng, W., Park, H.: Orthogonal nonnegative matrix t-factorizations for clustering. In: Proceedings of the 12th ACM SIGKDD International Conference on Knowledge Discovery and Data Mining, pp. 126–135. ACM (2006)
9. Sun, Y., Han, J., Yan, X., Yu, P.S., Wu, T.: Pathsim: Meta path-based top-k similarity search in heterogeneous information networks. In: VLDB 2011 (2011)

10. Sun, Y., Norick, B., Han, J., Yan, X., Yu, P.S., Yu, X.: Integrating meta-path selection with user-guided object clustering in heterogeneous information networks. In: Proceedings of the 18th ACM SIGKDD International Conference on Knowledge Discovery and Data Mining, pp. 1348–1356. ACM (2012)
11. Yu, X., Sun, Y., Norick, B., Mao, T., Han, J.: User guided entity similarity search using meta-path selection in heterogeneous information networks. In: Proceedings of the 21st ACM International Conference on Information and Knowledge Management, pp. 2025–2029. ACM (2012)
12. Dhillon, I.S.: Co-clustering documents and words using bipartite spectral graph partitioning: In: Proceedings of the Seventh ACM SIGKDD International Conference on Knowledge Discovery and Data Mining, pp. 269–274. ACM (2001)
13. Li, T., Ding, C., Zhang, Y., Shao, B.: Knowledge transformation from word space to document space. In: Proceedings of the 31st annual international ACM SIGIR Conference on Research and Development in Information Retrieval, pp. 187–194. ACM (2008)
14. Gu, Q., Zhou, J.: Co-clustering on manifolds. In: Proceedings of the 15th ACM SIGKDD International Cconference on Knowledge Discovery and Data Mining, pp. 359–368. ACM (2009)
15. Ding, C., Li, T., Jordan, M., et al.: Convex and semi-nonnegative matrix factorizations. IEEE Trans. Pattern Anal. Mach. Intell. **32**(1), 45–55 (2010)
16. Liu, J., Han, J.: Hinmf: A matrix factorization method for clustering in heterogeneous information networks. In: Proceedings of 2013 IJCAI Workshop on Heterogeneous Information Network Analysis (2013)
17. Liu, J., Wang, C., Gao, J., Gu, Q., Aggarwal, C., Kaplan, L., Han, J.: Gin: a clustering model for capturing dual heterogeneity in networked data. In: Proceedings of 2015 SIAM International Conference on Data Mining (2015)
18. Xu, W., Liu, X., Gong, Y.: Document clustering based on non-negative matrix factorization. In: Proceedings of the 26th Annual International ACM SIGIR Conference on Research and Development in Informaion Retrieval, pp. 267–273. ACM (2003)
19. Cai, D., He, X., Han, J., Member, S.: Document clustering using locality preserving indexing. IEEE Trans. Knowl. Data Eng. **17**, 1624–1637 (2005)
20. Lovsz, L., Plummer, M.: Matching Theory. Annals of Discrete Mathematics, vol. 29 inria-00345669, version 3 - 21 November 2009 (1986)

Data and Text Mining

A Profile-Based Authorship Attribution Approach to Forensic Identification in Chinese Online Messages

Jianbin Ma[1]($\boxtimes$), Bing Xue[2]($\boxtimes$), and Mengjie Zhang[2]

[1] College of Information Science and Technology,
Agricultural University of Hebei, Baoding 071001, China
majianbin@hebau.edu.cn
[2] School of Engineering and Computer Science,
Victoria University of Wellington, PO Box 600, Wellington 6140, New Zealand
{Bing.Xue,Mengjie.Zhang}@ecs.vuw.ac.nz

Abstract. With the popularity of Internet technologies and applications, inappropriate or illegal online messages have become a problem for the society. The goal of authorship attribution for anonymous online messages is to identify the authorship from a group of potential suspects for investigation identification. Most previous contributions focused on extracting various writing-style features and employing machine learning algorithms to identify the author. However, as far as Chinese online messages are concerned, they contain not only Chinese characters but also English characters, special symbols, emoticons, slang, etc. It is challenging for word segmentation techniques to segment Chinese online messages correctly. Moreover, online messages are usually short. The performance for short samples would be decreased greatly using traditional machine learning algorithms. In this paper, a profile-based authorship attribution approach for Chinese online messages is firstly provided. N-gram techniques are employed to extract frequency sequences, and the category frequency feature selection method is used to filter common frequent sequences. The profile-based method is used to represent the suspects as category profiles. The illegal messages are attributed to the most likely authorship by comparing the similarity between unknown illegal online messages and suspects' profiles. Experiments on BBS, Blog, and E-mail datasets show that the proposed profile-based authorship attribution approach can identify the authors effectively. Compared with two instance-based benchmark methods, the proposed profile-based method can obtain better authorship attribution results.

Keywords: Profile · Authorship attribution · N-gram · Chinese · Online messages · Forensic

1 Introduction

The number of Internet users in the world reach 3.2 billion in 2015 [19]. Nowadays, Internet is an important information source in people's daily life. People

© Springer International Publishing Switzerland 2016
M. Chau et al. (Eds.): PAISI 2016, LNCS 9650, pp. 33–52, 2016.
DOI: 10.1007/978-3-319-31863-9_3

can communicate by various mediums such as E-mail, BBS, Blog, and Chat Rooms. Unfortunately, these online communication mediums are being misused for inappropriate or illegal purposes. It is common to find out fraud information, antisocial information, terroristic threatening information, etc. [25]. Terrorists make use of Internet to post messages for radicalization and recruitment of youth. In China, there is an institute called "12321" in charge of inappropriate or junk information complaint and treatment. Taking the statistics from the "12321" institute in September 2015 as an example, the institute received complaints about 9,438 junk mails, 28,407 illegal websites, and 22,481 illegal or junk text messages on mobile phones [1]. So, the inappropriate or illegal online information has strongly disturbed people's daily life.

The obvious characteristics of cybercrime is anonymous and borderless. In order to escape from detecting, criminals always forge their personal information or send information anonymously [8]. Moreover, by the help of pervasive network technology, criminals can hide in any corder at any time to commit crime. So, it is hard to identify the real authorship of inappropriate or illegal web information.

Computer forensic has been used as evidence since the mid-1980s [6], which is a science for finding legal evidence in computers or digital storage medias. Forensic investigation methods try to dig various types of evidence for courts. Fingerprints, blood, hair, witness testimony, and shoe prints are the traditional incriminating types of digital investigation methods [14]. But the traditional forensic investigation methods are inapplicable to cybercrime investigation. There is limited information in the crime scenes. Only some electronic text messages are available. However, like footprint and handwriting, authors' writing-style features can be mined by analyzing the authors' writing habits on Internet medias.

The purpose of authorship attribution is to attribute the authorship of unknown writings according to writing-style analysis on the author's known works [25]. There are some typical studies such as authorship attribution on Shakespeare's works [12,26] and "The Federalist Papers" [17,27]. Since 2000, authorship attribution on E-mail, BBS and Chatting room information for forensic purpose have drew researchers' attention [3,8,11,20,35]. Current research focus on two aspects: feature extraction and selection methods, and authorship attribution algorithms. Identifying the authorship by analyzing the writing-style features from Chinese online messages is difficult due to complicated linguistic characteristics, which are analyzed as follows.

(1) Chinese language has no natural delimiters between words. The word segmentation is a key technique to process Chinese natural language. Nowadays, some word segmentation softwares are available. However, word segmentation softwares are difficult to segment correctly for some neologies such as slang on the online messages. The terms TMD and 886 are commonly used slang in Chinese online messages. It is difficult to segment such kinds of slang by word segmentation softwares. Further, Chinese online messages sometimes contain English words, which are important features to mine authors' writing-style features. So, feature extraction methods that are language independent need to be investigated and developed.

(2) Chinese online messages contain complicated linguistic elements, which include not only Chinese characters but also English characters, special symbols and emoticons. The authors usually write freely. There are a lot of extra blank spaces or blank lines. Moreover, there are symbols such as "∼∼∼", "....", and "!!!", emoticons such as ☺ and ☹, and english characters such as "byebye" and "bye" in Chinese online messages. These characters, symbols and emoticons can be extracted as writing-style features to identify the authorship of online messages. So, authorship attribution methods on Chinese online messages need more effective feature extraction and selection methods to make better use of the online message information.

(3) In general, online messages are short. The texts of online messages contain few words. The classification accuracy for short text would be decreased greatly in traditional machine learning algorithms, such as K-nearest neighbors (KNN) and support vector machines (SVMs) [7].

From the above analysis, we can see that authorship attribution methods on Chinese online messages should be investigated, which should be language independent, can analyze extensive linguistic elements including Chinese characters, English characters, digits, symbols, emoticons, and are suitable for dealing with short online message samples. Profile is used to represent the training texts per author. Stamatatos, et al. (2009) had compared the profile-based and instance-based methods [32], and described that the important advantages of the profile-based methods were that the profile-based methods might produce a more reliable representation when the training samples are short texts such as E-mail messages and online forum messages. Keselj et al. (2003) [23] and Peng et al. (2003) [28] presented authorship attribution methods based on character-level n-gram, which was language independent. So, in this paper, we employ n-gram techniques which is language independent to extract frequent sequences from extensive linguistic elements in Chinese online messages. The profile-based method is used to represent the suspects as category profiles.

One factor that influences authorship attribution accuracy is class imbalance. Class imbalance is caused by uneven distribution of the training samples over the candidate classes. Some candidate classes have more training samples, while some other classes have fewer training samples. Moreover, the text lengths of training samples are different. Some candidate classes have longer texts than other candidate classes. Most previous authorship attribution approaches work well when the training samples are balanced, namely, equal number of training samples for each candidate class and the almost same length of training samples for each candidate calss. However, in most cases, the training samples over the candidate classes are not balanced. In such a situation, most of classifiers are biased toward the majority class [22]. Class imbalance problem will greatly reduce authorship attribution accuracy.

Class imbalance can cause frequent sequence imbalance in profiles. The number of frequent sequences in majority classes is more than that of minority classes. Further, the feature value of the frequent sequence in majority classes is larger than that of minority classes based on the n-gram based frequent sequence extraction and selection method (shown in Sect. 3.2). So, in this paper, a frequent sequence standardization method is introduced to solve class imbalance problem.

1.1 Goals

In this paper, the overall goal is to propose a profile-based authorship attribution approach for Chinese online messages to effectively identify the author from a list of suspects and provide convictive evidence for cybercrime investigation. We will focus on the following four objectives in order to achieve the overall goal.

Objective 1: Employ n-gram techniques to extract frequent sequences from extensive linguistic elements including Chinese characters, English characters, digits, symbols, etc., and use the category frequency features selection method to filter common frequent sequences that do not have distinguished ability.

Objective 2: Develop a profile-based method to represent the suspects to category profiles, and present a similarity computing method to attribute unknown illegal messages to most likely authorship.

Objective 3: Develop a frequent sequence standardization method to solve class imbalance problem, and investigate whether the method is effective.

Objective 4: Propose a profile-based authorship attribution approach based on the above methods, and investigate whether this approach can obtain effective experimental results, and achieve better performance than two instance-based benchmark methods.

1.2 Organization

The rest of the paper is organized as follows. Section 2 reviews the previous contributions. Section 3 presents the proposed profile-based authorship attribution method. Section 4 describes the experiment design. Section 5 provides the experimental results and discussions. Section 6 are the conclusions and future work.

2 Related Works

Authorship attribution was used to attribute the authorship of literatures. The pioneering authorship attribution methods traced back to Mosteller and Wallace (1964) [27] who tried to attribute the authorship of "The Federalist Papers". Since the late 1990s, the vast amount of electronic texts (E-mail, Blog, Online forum, etc.) have appeared on the Internet. Authorship attribution studies have been used on forensic investigation [2,8,11,20]. In this section, we review n-grams writing-style features, authorship attribution method for forensic identification, and two typical writing-style feature representation methods.

2.1 N-Grams Writing-Style Features

In the views of computational linguistics, an n-gram is a contiguous sequence of n items extracted from a sequence of text. N-grams at the character-level were widely applied to authorship attribution. There were some successful applications. Forsyth and Homes (1996) [15] found that bigrams and charater n-grams achieved better performance than lexical features in authorship attribution. Keselj et al. (2003) [23] and Peng et al. (2003) [28] presented an authorship attribution method using character-level n-gram, which was language independent. Sun et al. (2012) [33] proposed an online writeprint identification framework using variable length character n-gram to represent the author's writing-style. The items of an n-gram can be characters and words according to the previous application. However, more extensive linguistic elements in Chinese online messages including Chinese characters, English characters, digits, symbols, emoticons should be extracted. In this paper, these linguistic elements are termed n-gram frequent sequences.

2.2 Authorship Attribution Method for Forensic Investigation

De (2000, 2001) [8–10] extracted a set of linguistic and structural features, and employed SVMs algorithm to attribute the authorship of E-mail documents. Iqbal et al. (2008, 2010) [20,21] mined frequent patterns for authorship attribution in E-mail forensic investigation. However, in the pre-processing phase, the spaces, punctuations, special characters and blank lines which are important information that can be used to mine authors' writing-styles are removed. Zheng et al. (2003, 2006) [35,36] presented an authorship attribution method that extracted a comprehensive set of syntactical features, lexical features, structural features, and content-specific features, and used inductive learning algorithms to build classification models. Chen (2008) [3] extracted a rich set of writing-style features and developed the Writeprints technique for identification and similarity detection on online messages. Ding et al. (2015) [11] proposed a visualizable evidence-driven approach based on an End-to-End Digital Investigation framework to visualize and corroborate the linguistic evidence supporting output attribution results.

2.3 Writing-Style Feature Representation Methods

The purpose of authorship attribution techniques is to form an attribution model by analyzing the writing-style features from the training corpus. Then, the attribution model attributes text samples of unknown authorship to a candidate author. The authorship attribution methods can be divided into two classes, namely, profile-based methods and instance-based methods. Profile-based methods extract a general style (called the author's profile) for each author from available training texts. Instance-based methods treat each text sample in the training set as an instance and extract a separate style for each text sample. Classification algorithms are often used to develop an attribution model in instance-based methods.

Profile-Based Authorship Attribution Methods. Keselj et al. (2003) [23] proposed an authorship attribution method that used n-grams to form author profiles. Iqbal et al. (2008) [21] extracted author profiles called write-prints based on frequent patterns extraction method. Estival et al. (2007) [13] presented an author profiling method with the application to Arabic E-mail authorship attribution. Stamatatos (2009) [32] had compared the profile-based and instance-based methods. He described that the important advantage of profile-based methods were that the profile-based methods might produce a more reliable representation when only short texts were available for training such as E-mail messages and online forum messages. Further, the computation time of profile-based methods was lower than that of instance-based methods.

Instance-Based Authorship Attribution Methods. In instance-based methods, machine learning algorithms such as neural networks, Bayesian, decision trees, KNN and SVMs were widely used to train an attribution model. Merriam and Matthews (1994) [26] employed neural network classifiers in the authorship attribution study. Kjell (1994a) [24] employed Bayesian and neural networks as classifiers. Hoorn et al. (1999) [18] extracted letter sequences for authorship analysis of three Dutch poets using neural networks. Holmes (1998) [16] used a genetic rule based learner for "The Federalist papers" authorship attribution problem by comparing the effects of vocabulary richness, and word frequency analysis.

Most authorship attribution studies focus on various pre-defined writing-style features and test different feature sets on effect of experimental results. Due to Chinese online messages' complicated linguistic characteristics (shown in Sect. 1), effective feature extraction, selection, and representation methods that are suitable for Chinese online messages authorship attribution should be investigated. Keselj et al. (2003) [23] and Peng et al. (2003) [28] proposed a language independent authorship attribution method using character-level n-gram language models. However, the method is restricted to character level, and merely applied to literature documents. They did not consider integrating extensive linguistic elements including Chinese characters, English characters, digits, symbols, emoticons, etc. So, in this paper, we propose a profile-based authorship attribution approach for Chinese online messages. Frequent sequences that are combinations of syntactic features, lexical features, and structural features are extracted by n-gram techniques from extensive linguistic elements. The frequent sequences are represented as category profiles. A similarity computing method is employed to attribute unknown illegal messages to most likely authorship.

3   The Proposed Profile-Based Authorship Attribution Approach

3.1   Overview

The process of the proposed profile-based authorship attribution approach for Chinese online messages, as shown in Fig. 1, can be divided into the following five steps.

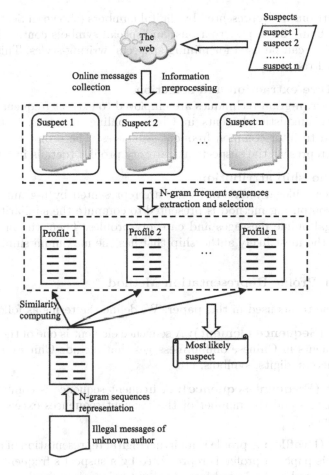

Fig. 1. The process of profile-based authorship attribution approach for Chinese online messages

Step 1. Preliminary investigation

Let's suppose there are illegal online messages that cause bad effect, and the Police Department intends to investigate who wrote these messages on the Internet. By preliminary investigation, investigators could narrow down the suspect list.

Step 2. Online messages collection

Each suspect's online message samples should be collected. These samples are suspect's known works. The goal here is to collect samples as many as possible. These online message samples are used as the training set.

Step 3. Online messages pre-processing

Some useless information such as advertisements, pictures, multimedia information should be removed. Text information and emoticons are reserved. In our research, emoticons are represented by digital numbers (shown in Sect. 3.2). The blank spaces, blank lines, punctuations, and special symbols contain important information that can be used for mining suspects' writing-styles. This information is reserved too.

Step 4. Feature extraction and selection

In this paper, n-gram techniques are employed to extract frequent sequences from extensive linguistic elements in Chinese online messages. Category frequency is used to filter common frequent sequences. The profile-based method is adopted to represent the suspects as category profiles (details in Sect. 3.2).

Step 5. Authorship attribution

The unknown illegal online messages are represented by n-gram sequences. A similarity computing method is presented to compute the similarity between unknown illegal online messages and category profiles. Then, the messages are attributed to the most likely authorship that has the maximum similarity.

3.2   Author Profile Representation Method

There are some terms used in this paper. We define the terms as follows.

Definition 1 (Sequence element): A sequence element is one of the minimum linguistic elements in Chinese online messages that include Chinese characters, English characters, digits, symbols, etc.

Definition 2 (Frequent sequence): A frequent sequence is combinations of sequence elements and the number of the frequent sequences exceed the given support threshold.

Definition 3 (Profile): A profile is a virtual digital representation of a suspect's identity. In this paper, a profile is represented by a suspect's frequent sequences which are extracted and selected from the suspect's training set.

Let us suppose there is a suspect list $S = \{S_1, S_2, \cdots, S_n\}$, n is the number of suspects. There is an illegal message d that the author is unknown. The decision function is thus asked to map newly incoming illegal message d in one suspect from the suspect list (S), according to its content.

Profile-based classifiers derive a description of each target class (S_i) in terms of a category profile (C_i), usually a vector of features. These vectors are extracted from a training set $D = \{(d_1, y_1), (d_2, y_2), \cdots, (d_m, y_m)\}$ that is pre-categorized under C_i, where d_i is the feature vector of the ith sample, y_i is its label (i.e., category), and m is the number of samples. The profile-based classifiers can be referred as *category-centered* classification, which is thus the evaluation of similarity between unknown document d and different profiles (one for each class) [5].

Given a set of features $\{t_1, t_2, \cdots, t_u\}$ describing an online message $d_h \in D$, u is the number of features. The online message is represented as a feature vector, i.e.: $d_h = \{(t_1, w_{t_1}^h), (t_2, w_{t_2}^h), \cdots, (t_u, w_{t_u}^h)\}$, where $w_{t_k}^h$ represents the feature value t_k for online message d_h. Category profiles of suspects are represented as vectors of features, i.e.: $C_i = \left\{ \left(t_1, w_{t_1}^i\right), \left(t_2, w_{t_2}^i\right), \cdots, \left(t_u, w_{t_u}^i\right)\right\}$, where $w_{t_k}^i$ represents the feature value t_k in category profile C_i. In this paper, n-gram techniques are employed to extract frequent sequences as the writing-style features.

N-Gram Based Frequent Sequence Extraction and Selection Method.
Each online message consists of a sequence of characters. The characters include Chinese characters, English characters, digits, punctuations, blank lines, blank spaces, special symbols, emoticons etc.

Based on the following rules, the sequence of characters is combined into sequence elements.

Rule 1: if the element of a sequence is a Chinese character, then the Chinese character is a sequence element.

Rule 2: if the element of a sequence is an English character, then read the next element. All the English characters are treated as a single sequence element until the element is not English character.

Rule 3: if the element of a sequence is a punctuation, then read the next elements. All the punctuations are treated as a single sequence element until the element is not punctuation.

Likewise, the elements of digits, blank spaces, blank lines, tab spaces are suitable for the rules 2 and 3.

An emoticon is represented as three-digit numbers such as <001> or <002>. The representation of an emoticon is treated as a sequence element.

Thus, based on the above rules, the extensive linguistic elements for Chinese online messages is transformed to sequence elements.

Then, the sequence elements are combined to 1-gram sequences, 2-gram sequences, 3-gram sequences, 4-gram sequences, etc. In the experiment section, experimental results show that 4-gram sequences are more effective than other n-grams (details in Sect. 5).

Most n-gram sequences occur in a certain category only once. These features are rare features which are either noninformative for category prediction or not influential in global performance [34]. We define frequent features that occur in the unique category frequently. The features are treated as frequent features on condition that the frequency of features exceeds the given frequent support threshold defined as "Minimum Term Frequency, *TFmin*", which is the smallest number of times a feature can appear in a category. The sequences that meet the condition of *TFmin* are extracted as features to remove the rare features.

If most category profiles containing one certain frequent sequence, the frequent sequence does not have distinguished ability. Filtering common frequent sequences among profiles and selecting effective frequent sequences are essential. We define the distinguish ability of one frequent sequence as category frequency (*CF*), which is the number of category profiles in which a frequent sequence occurs. We computed the *CF* for each unique frequent sequence in the training set and removed from the search space those frequent sequence whose *CF* exceed the given minimum category frequency threshold ($CFmin$). Let us suppose all the unique frequent sequences among category profiles are represented as $T = \{(t_1, e_1), (t_2, e_2), \cdots, (t_v, e_v)\}$, where t_i is the ith frequent sequence, e_i is the total number of category profiles that contain the frequent sequence t_i, v is the number of unique frequent sequences. Formula (1) is the condition of filtering common frequent sequences.

$$\frac{e_i}{n} >= CFmin \tag{1}$$

where n is the number of profiles. The frequent sequences that satisfy the Formula (1) are removed.

The usual TF-IDF scheme is widely used in the vector space model [30]. In this paper, we improve the TF-IDF method and propose a new TFC-ICF method to compute the feature value of frequent sequences in the category profiles. TFC represents the number of a frequent sequence that occurs in the online messages of a certain category. The inverse category frequency (ICF) [4] is similar to the inverse term frequency (IDF) [31]. ICF is given by $ICF = \log\left(\frac{N}{F_{t_k}}\right)$, where F_{t_k} is the number of category profiles in which a frequent sequence $t_k \in \{t_1, t_2, \cdots, t_u\}$ occurs, N is the total number of categories. Then, the category profile feature values ($w_{t_k}^i$) are computed by the formula (2).

$$w_{t_k}^i = TFC \times ICF = \sum_{h \in C_i} G_{t_k}^h \times \log\left(\frac{N}{F_{t_k}}\right) \tag{2}$$

where $G_{t_k}^h$ is the number of frequent sequence t_k that occurs in the online message d_h. C_i is the ith category profile.

Solving Class Imbalance Problems. In this paper, a frequent sequence standardization method to solve class imbalance problem is introduced. In category profile list $C = \{C_1, C_2, \cdots, C_n\}$, non-zero frequent sequences in each profile $C_i = \{(t_1, w_{t_1}^i), (t_2, w_{t_2}^i), \cdots, (t_u, w_{t_u}^i)\}$ are sorted by the feature values in descending order. Then, every non-zero frequent sequence has its own ranking in the category profile. The total number of non-zero frequent sequences in each profile is counted. Let us suppose category profile C_i has the smallest number of non-zero frequent sequences, and the number is r. The number of non-zero frequent sequences in each profile is kept down to the same number r according to their ranking. The feature values of the rest of low ranking (lower than r) frequent sequences in profiles is set to 0. The feature values of non-zero frequent sequences in each profile are normalized to the range between 0 and 1.

The feature value is computed by the Formula (3).

$$w_{t_k}^i = \frac{r - s}{r - 1} \tag{3}$$

where $w_{t_k}^i$ is the feature value of non-zero frequent sequence t_k in the category profile C_i. s is the ranking of the non-zero frequent sequence t_k. r is the total number of the non-zero frequent sequences in profile C_i. $\frac{1}{r-1}$ is the interval between two non-zero frequent sequences.

3.3   Authorship Attribution

Let us suppose there is an illegal online message d. The message d is represented to character sequences, sequence elements and n-gram sequences. Ultimately, the message d is represented to $d = \{(t_1, w_{t_1}), (t_2, w_{t_2}), \cdots, (t_u, w_{t_u})\}$, where $w_{t_i} = ICF \times G_{t_i}$, The computation formula of ICF is shown in Sect. 3.2, G_{t_i} is the number of feature t_i occurs in the message d. Formula (4) is the similarity function between unknown document d and category profile C_i.

$$sim(d, C_i) = \cos(\boldsymbol{d}, \boldsymbol{C_i}) = \sum_{k=1}^{u} \frac{w_{t_k} \cdot w_{t_k}^i}{|\boldsymbol{d}| \cdot |\boldsymbol{C_i}|} \tag{4}$$

where $sim(d, C_i)$ is the similarity degree between unknown online message d and category profile C_i, w_{t_k} is the feature value of sequence t_k in the message d, $w_{t_k}^i$ is the feature value of frequent sequence t_k in category profile C_i, u is the number of frequent sequences.

The most likely category of an unknown online message d is computed by Formula (5).

$$author\,(d) = \arg\max sim(d, C_i) \tag{5}$$

The unknown online message d is attributed to the maximum similarity between the unknown online message d and profile $C_i \in C$.

4   Experiment Design

In this section, experiments are designed to evaluate the performance of the proposed profile-based authorship attribution approach. The overall experimental objective is to verify whether our method can effectively identify the author from a list of suspects. Three experiments are performed to test the overall experimental objectives. (1) The first experiment is to test the influence of different parameters *TFmin* and *CFmin* on experimental results. (2) The second experiment is to test whether the accuracy is influenced by the class imbalance problem obviously, and show whether our frequent sequence standardization method is effective. (3) The third experiment is to compare the experimental results of the proposed profile-based approach and two benchmark methods, and test whether the proposed profile-based approach can achieve better performance than the two benchmarks.

4.1   Datasets

Three real-life datasets including BBS, Blog, and E-mail were collected. Table 1 shows the detailed information of the three datasets. There is no public Chinese datasets for online messages. So, we collected 10 most popular Bloggers on the website http://blog.sina.com.cn/ as the blog dataset. BBS dataset were gained from 6 active moderators on the Zhihu web forum http://www.zhihu.com/. Involving personal privacy, E-mails were collected from 5 staff members' mailboxes that were used to announce notifications on our university's mail server.

Table 1. The information of three datasets

Dataset	Authors	No. of instances	Average no. of words per instance
Blog	Author 1	200	273
	Author 2	200	926
	Author 3	200	354
	Author 4	107	336
	Author 5	200	413
	Author 6	200	461
	Author 7	200	184
	Author 8	183	289
	Author 9	197	610
	Author 10	200	227
BBS	Author 1	68	78
	Author 2	117	40
	Author 3	92	34
	Author 4	79	25
	Author 5	90	16
	Author 6	90	65
E-mail	Author 1	22	479
	Author 2	28	194
	Author 3	19	171
	Author 4	10	320
	Author 5	19	168

From the information of the three datasets in Table 1, we can see that the Blog dataset has more instances and the average number of words per instance is longer than other datasets. The average number of words per instance in the BBS dataset is fewer than that of other datasets. The class imbalanced problem in BBS and E-mail datasets are more obvious than Blog dataset.

4.2   Benchmarks for Comparison

To test the effectiveness of the proposed profile-based authorship attribution approach, two instance-based methods were selected as benchmarks for comparison.

There were little related studies aiming at authorship attribution method for Chinese online messages on forensic purpose except for our artificial intelligence and data mining research group in agricultural university of Hebei. So, we selected our previous instance-based authorship attribution method [25] as the first benchmark. We term the first benchmark as *Instance-lexi-stru* method. In the *Instance-lexi-stru* method, word segmentation softwares were used for word segmentation and part of speech tagging due to Chinese language's special characteristics. The information gain method was used to select effective lexical features. The feature value of lexical features was calculated by the traditional *TF-IDF* Formula [29].

$$w(t, \boldsymbol{d}) = tf(t, \boldsymbol{d}) \times log(N/n_t + 0.01) \qquad (6)$$

where $w(t, \boldsymbol{d})$ is the feature value of feature t in document d, $tf(t, \boldsymbol{d})$ is the frequency of feature t in document d, N is the total number of documents, and n_t is the number of documents that contain feature t.

Structural features include structural characteristics (shown in Table 2) [25], punctuations features (30 categories including Chinese and English punctuations), and part of speech features (12 categories part of speech features). An SVM are used as learning algorithm.

Table 2. Structural characteristics

Features
Number of distinct punctuations/total number of punctuations
Number of distinct words/total number of words
Mean sentence length
Mean paragraph length
Number of digital characters/total number of words
Number of lowercase letters/total number of words
Number of uppercase letters/total number of words
Number of space/total number of words
Number of blank lines/total number of lines
Number of indents/total number of words

In the second benchmark, we employed n-gram techniques to extract n-gram frequent elements (detail shown in Sect. 3.2). The information gain method was used to select effective n-gram frequent elements. The *TF-IDF* in Formula (6) is used to compute the feature value of n-gram frequent elements. An SVMs are used as learning algorithm. We term the second benchmark as *instance-n-gram* method.

In the experiments, the samples in each dataset were randomly divided into two sets, namely, 70 % as the training set, and 30 % as the test set. The accuracy was used to evaluate the experimental results.

5   Experimental Results and Discussions

5.1   The Influence of *TFmin* and *CFmin* on Experimental Results

To test the influence of different parameters *TFmin* and *CFmin* on experimental results, the first experiment was conducted. Different parameter combinations were experimented on the three datasets to find effective parameters combinations. The experimental results were shown in Table 3.

Table 3. The experimental results of different parameter combinations

Dataset	Parameters combinations		Accuracy (%)			
	TFmin	*CFmin*	2-gram	3-gram	4-gram	5-gram
Blog	2	0.5	84.18	87.03	87.50	86.87
		0.6	83.86	87.50	87.82	87.18
		0.7	85.76	87.82	88.13	87.34
		0.8	85.76	88.13	**88.29**	8.66
		0.9	85.92	87.82	88.13	88.13
	3	0.5	81.65	84.34	85.29	85.13
		0.6	82.60	85.44	85.44	85.13
		0.7	82.44	86.23	87.76	86.08
		0.8	82.75	84.97	86.23	86.08
		0.9	83.70	84.97	87.76	85.60
	4	0.5	80.22	82.91	82.91	82.75
		0.6	80.06	84.02	83.86	83.86
		0.7	80.54	83.70	84.81	84.34
		0.8	81.96	84.45	85.60	86.08
		0.9	81.65	84.97	85.13	85.13
BBS	2	0.5	69.60	70.64	69.80	69.05
		0.7	73.68	74.44	**75.39**	75.19
		0.9	69.63	71.85	70.37	71.11
	3	0.5	67.48	67.74	69.05	65.08
		0.7	59.56	59.26	60.45	61.19
		0.9	55.97	55.97	58.96	58.21
	4	0.5	0	0	0	0
		0.7	52.73	51.15	55.00	55.00
		0.9	41.84	42.17	43.02	47.71
E-mail	2	0.4	77.42	77.42	77.42	77.42
		0.6	80.65	80.65	**83.87**	80.65
		0.8	80.66	80.66	77.42	77.42
	3	0.4	55.00	61.62	77.78	73.42
		0.6	61.91	55.00	70.22	66.67
		0.8	56.57	64.29	64.29	70.74
	4	0.4	0	57.94	65.07	65.52
		0.6	53.15	53.15	65.00	61.67
		0.8	56.57	56.57	62.26	53.00

From Table 3, we can see that the experimental results of the 4-gram column are better than those of the 2-gram, 3-gram and the 5-gram columns in most cases, which suggests that the 4-gram frequent sequences is more effective than other n-grams. With regard to parameter *TFmin*, the experimental results on the condition that the parameter value is 2 are better than those of other parameter values, which suggests that the sequences are frequent sequences when at least 2 documents in a certain category contain the sequences. As for parameter *CFmin*, the highest accuracy on the Blog dataset is 88.29 % on the condition that the parameter value of *CFmin* is 0.8, and the parameter value of *TFmin* is 2. There is little difference among the experimental results on blog dataset when the parameter value of *CFmin* is 0.7, 0.8 and 0.9. Likewise, the highest experimental result on the BBS dataset is 75.39 % on the condition that the parameter value of *CFmin* is 0.7, and the parameter value of *TFmin* is 2. The highest experimental result on the E-mail dataset is 83.87 % on the condition that the parameter value of *CFmin* is 0.6, and the parameter value of *TFmin* is 2. If the parameter value of *CFmin* is too high, the purpose to filter common frequent sequence would not be achieved. If the parameter value of *CFmin* is too low, some frequent sequences that have distinguished ability would be filtered away. So, the appropriate parameter value of *CFmin* should be set at a range of 0.6 to 0.8.

From Table 3, we can see that highest accuracy of Blog, BBS, and E-mail are 88.29 %, 75.39 %, and 83.87 % respectively. The accuracy exceeds 80 % by experimenting on the Blog and E-mail datasets. The accuracy of the E-mail dataset is high, because E-mail documents have obvious structural features such as greetings, farewells, and signatures, and the n-gram based frequent sequences feature extraction and selection method can represent the E-mail author's writing-styles well. The accuracy of the BBS dataset is relatively low, which might be caused by too few words in the BBS test set. Some BBS test samples even can not find the matching frequent sequences in any category profiles. The writing-styles in too short online messages are not obvious. Experimental results show that the authors can be identified from a list of suspects effectively, and promising performance is achieved.

5.2   The Experiments on Class Imbalance Problem

In this paper, a frequent sequence standardization method is used to deal with the class imbalance problem. To test the effectiveness of the frequent sequence standardization method, experiments were done on the Blog dataset. We changed the number of instances for authors randomly to produce the class imbalance problem. Firstly, the two authors' instances as shown in Table 1 were reduced by half. Then, the four authors' instances were reduced by half, and so on. The experimental results are shown in Table 4. We assume the parameter η denote the proportion of the number of authors whose instances are reduced by half in all the authors.

From the experimental results in Table 4, we can see that the accuracy does not change obviously. The accuracy decreases slightly when the two authors' instances are reduced by half. Then, the accuracy keeps stable when more authors' instances are reduced by half, which suggests that class imbalance problem has little effect on experimental performance and shows that the frequent sequence standardization method to solve class imbalance problem is effective.

Table 4. The experimental results of class imbalance situation for the Blog dataset

η	Accuracy (%)
0/10	88.29
2/10	85.76
4/10	84.65
6/10	84.86
8/10	85.13
10/10	84.49

5.3 Comparison of the Proposed Profile-Based Method with Two Benchmark Methods

In Sect. 4.2, two instance-based benchmarks (*Instance-lexi-stru* and *instance-n-gram*) are described. Each sample of the training set is treated as an instance and is extracted to a separate writing-style. The method is different from the proposed profile-based approach in this paper. We term the proposed profile-based approach as *Profile-n-gram*. To compare the performance of *Profile-n-gram*, *Instance-lexi-stru* and *Instance-n-gram*, experiments were made on three datasets. In instance-based methods, the kernel function was set to the linear kernel function in SVMs. 1000 lexical features were selected by the information gain features selection method. In *Profile-n-gram*, the parameters of *TFmin* and *CFmin* were set to optimal parameter combination based on the Sect. 5.1. The experimental results on different datasets are shown in Table 5.

From Table 5, we can see that the accuracy of the proposed *Profile-n-gram* method on each dataset is higher than that of the *Instance-lexi-stru* and *Instance-n-gram* methods. The accuracy of the *Instance-lexi-stru* on Blog and BBS datasets is higher than that of the *Instance-n-gram*. The accuracy of the

Table 5. The experimental results of *Profile-n-gram*, *Instance-lexi-stru* and *Instance-n-gram*

Dataset	Method type	Accuracy (%)
Blog	Profile-n-gram	88.29
	Instance-lexi-stru	82.06
	Instance-n-gram	80.15
BBS	Profile-n-gram	75.39
	Instance-lexi-stru	62.94
	Instance-n-gram	61.24
E-mail	Profile-n-gram	83.87
	Instance-lexi-stru	72.74
	Instance-n-gram	80.66

Instance-n-gram on the E-mail dataset is higher than that of the *Instance-lexi-stru*, because the E-mail dataset has obvious structural features, and n-gram feature extraction and selection method can mine the writing-style features effectively. The accuracy gap between the profile-based and the instance-based methods on the BBS dataset is wider than that of the Blog dataset, which might be caused by the characteristic of the two datasets. From Table 1, we can see that the length of samples in the Blog dataset is longer than that of the BBS dataset. The short samples would decrease the performance of the instance-based methods greatly [7]. The proposed profile-based methods can take full advantages of short online messages, and extract frequent sequences that almost can not be mined by traditional feature extraction methods [3, 20, 35].

Examples of frequent sequence extraction results for profiles on the BBS dataset is shown in Table 6. The left side of "=" is the frequent sequence. The right side of "=" is the feature value of the frequent sequence. Some fixed collocations in author's profiles include the combinations of Chinese words, English words, punctuations, modal words, digits, and slang words. The experimental results show that the n-gram based frequent sequence feature extraction methods are suitable for Chinese online message characteristics, and the proposed profile-based method is effective for solving authorship attribution problems with short online message samples.

Table 6. Examples of frequent patterns for profiles on BBS dataset

Authors	Examples of frequent patterns for different profiles
Author 1	大叔=0.91, 吗?=0.88, 呵呵..=0.75, 嘿嘿..=0.63, 了?=0.63 了 =0.55, 呵呵 =0.34, 阿邦=0.34, 拜拜=0.34, 嫣然=0.34
Author 2	???=0.94, ...=0.92, 苹果=0.81, 了。=0.72, 的。=0.72 我们=0.64, 但=0.56, 球员=0.42, 这=0.42, iphone=0.19
Author 3	。=0.97, ?=0.91, ...=0.92,如果=0.67, 中国=0.56 然后=0.27,时候=0.34, 国民=0.27, 可能=0.27, 微博=0.15
Author 4	呵呵=0.92, 了....=0.86, 啊 =0.8, ~~~=0.72, 哈哈=0.72 这=0.72, 就=0.55, 挺好的=0.47, 嗯的=0.47, 空看=0.47
Author 5	~~~=0.97, 我=0.94, 呵呵=0.92, 帖子=0.89, 大哥=0.84 吧主=0.73, 哈哈=0.64, 不过=0.58, 真的=0.47, 谢谢!=0.19
Author 6	......=0.98, bye-bye=0.86, hi=0.76, 但是=0.67, 游戏=0.59 不能=0.45, 还是=0.45, 科技=0.45, 开始=0.34, 那个=0.25

6 Conclusions

In this paper, a profile-based approach to authorship attribution for Chinese online messages was firstly proposed. N-gram techniques were employed to extract frequent sequence from extensive linguistic elements in Chinese online messages, and category frequency was used to filter common frequent sequences. A novel frequent sequence standardization method was used to deal with class imbalance problem. The profile method was used to represent the authors as category profiles. A similarity computing method was employed to attribute illegal online messages to the most likely authorship.

To test the effectiveness of the proposed profile-based authorship attribution approach, Blog, BBS and E-mail datasets were collected. Experimental results showed that highest accuracy of Blog, BBS, and E-mail were 88.29 %, 75.39 %, and 83.87 %, respectively. Comparing with the two benchmark methods, the proposed profile-based authorship attribution approach achieved better performance than the two instance-based benchmark methods. The frequent sequence standardization method to solve class imbalanced problem was effective. This study showed that the proposed profile-based authorship attribution approach for Chinese online messages could identify the author from a list of suspects effectively, and present convictive evidence for cybercrime investigation.

Our future research will focus on the following directions. First, we intend to extend the writing-style features by semantical analysis on the training samples based on Chinese linguistic characteristics. Second, no standard benchmark datasets are currently available. To achieve more meaningful experimental results, we will try to collect more real-world datasets and experiment on those datasets. Third, we will investigate cybercrime forensic methods that combine our authorship attribution methods with other specific forensic technologies such as data recovery, tracing IP address, tracing log file, and social relation network investigation.

Acknowledgments. This work was supported by grants from Department of Education of Hebei Province(No.QN20131150), Program of Study Abroad for Young Teachers by Agricultural University of Hebei. The authors also gratefully acknowledge the helpful comments and suggestions of the reviewers, which have improved the presentation.

References

1. 12321: 12321 statistics figures (2015). http://12321.cn/report.php
2. Abbasi, A., Chen, H.: Applying authorship analysis to extremist-group web forum messages. IEEE Intell. Syst. **20**(5), 67–75 (2006)
3. Abbasi, A., Chen, H.: Writeprints: a stylometric approach to identity-level identification and similarity detection in cyberspace. ACM Trans. Inf. Syst. (TOIS) **26**(2), 1–29 (2008)
4. Basili, R., Moschitti, A., Pazienza, M.T.: A text classifier based on linguistic processing. In: Proceedings of IJCAI99, Machine Learning for Information Filtering. Citeseer, Stockholm, Sweden (1999)
5. Basili, R., Moschitti, A., Pazienza, M.T.: Robust inference method for profile-based text classification. In: Proceedings of JADT 2000, 5th International Conference on Statistical Analysis of Textual Data. Lausanne, Switzerland (2000)
6. Casey, E.: Digital Evidence and Computer Crime: Forensic science, Computers, and the Internet. Academic press, Cambridge (2011)

7. Chen, M., Jin, X., Shen, D.: Short text classification improved by learning multi-granularity topics. In: Proceedings of the 22nd International Joint Conference on Artificial Intelligence, pp. 1776–1781. Citeseer, Barcelona, Spain (2011)
8. De Vel, O.: Mining e-mail authorship. In: Proceedings of ACM International Conference on Knowledge Discovery and Data Mining (KDD 2000). Boston, USA (2000)
9. De Vel, O., Anderson, A., Corney, M., Mohay, G.: Mining e-mail content for author identification forensics. ACM SIGMOD Rec. **30**(4), 55–64 (2001)
10. De Vel, O., Anderson, A., Corney, M., Mohay, G.: Multi-topic e-mail authorship attribution forensics. In: Proceedings of ACM Conference on Computer Security - Workshop on Data Mining for Security Applications. ACM, Philadelphia, PA, USA (2001)
11. Ding, S.H.H., Fung, B.C.M., Debbabi, M.: A visualizable evidence-driven approach for authorship attribution. ACM Trans. Inf. Syst. Secur. (TISSEC) **17**(3), 12 (2015)
12. Elliot, W., Valenza, R.: Was the earl of oxford the true shakespeare. Notes Queries **38**(4), 501–506 (1991)
13. Estival, D., Gaustad, T., Pham, S.B., Radford, W., Hutchinson, B.: Tat: an author profiling tool with application to arabic emails. In: Proceedings of the Australasian Language Technology Workshop, Melbourne, Australia, pp. 21–30 (2007)
14. Fisher, B.A., Fisher, D.R.: Techniques of Crime Scene Investigation. CRC Press, Boca Raton (2012)
15. Forsyth, R.S., Holmes, D.I.: Feature-finding for text classification. Literary Linguist. Comput. **11**(4), 163–174 (1996)
16. Holmes, D.I.: The evolution of stylometry in humanities scholarship. Literary Linguist. Comput. **13**(3), 111–117 (1998)
17. Holmes, D.I., Forsyth, R.S.: The federalist revisited: new directions in authorship attribution. Literary Linguist. Comput. **10**(2), 111–127 (1995)
18. Hoorn, J.F., Frank, S.L., Kowalczyk, W., van Der Ham, F.: Neural network identification of poets using letter sequences. Literary Linguist. Comput. **14**(3), 311–338 (1999)
19. ICT: Ict facts and figures (2015). http://www.itu.int/en/ITU-D/Statistics/Pages/facts/default.aspx
20. Iqbal, F., Binsalleeh, H., Fung, B.C., Debbabi, M.: Mining writeprints from anonymous e-mails for forensic investigation. Digit. Invest. **7**(1), 56–64 (2010)
21. Iqbal, F., Hadjidj, R., Fung, B.C.M., Debbabi, M.: A novel approach of mining write-prints for authorship attribution in e-mail forensics. Digit. Invest. **5**, S42–S51 (2008)
22. Japkowicz, N., Stephen, S.: The class imbalance problem: a systematic study. Intell. Data Anal. **6**(5), 429–449 (2002)
23. Kešelj, V., Peng, F., Cercone, N., Thomas, C.: N-gram-based author profiles for authorship attribution. In: Proceedings of the Conference Pacific Association for Computational Linguistics, PACLING, vol. 3, pp. 255–264. Halifax Canada, (2003)
24. Kjell, B.: Authorship attribution of text samples using neural networks and Bayesian classifiers. In: Proceedings of IEEE International Conference on Systems. Man, and Cybernetics, vol. 2, pp. 1660–1664. IEEE, San Antonio, USA (1994)
25. Ma, J.B., Li, Y., Teng, G.F.: CWAAP: an authorship attribution forensic platform for chinese web information. J. Softw. **9**(1), 11–19 (2014)
26. Merriam, T.V., Matthews, R.A.: Neural computation in stylometry II: an application to the works of Shakespeare and Marlowe. Literary Linguist. Comput. **9**(1), 1–6 (1994)

27. Mosteller, F., Wallace, D.: Inference and Disputed Authorship: The Federalist. Addison-Wesley, Boston (1964)
28. Peng, F., Schuurmans, D., Wang, S., Keselj, V.: Language independent authorship attribution using character level language models. In: Proceedings of the tenth conference on European chapter of the Association for Computational Linguistics. vol. 1, pp. 267–274. Association for Computational Linguistics, Stroudsburg, USA (2003)
29. Rajaraman, A., Ullman, J.D.: Mining of Massive Datasets, vol. 77. Cambridge University Press, Cambridge (2011)
30. Salton, G., Buckley, C.: Term-weighting approaches in automatic text retrieval. Inf. Process. Manag. **24**(5), 513–523 (1988)
31. Sichel, H.S.: On a distribution law for word frequencies. J. Am. Stat. Assoc. **70**(351a), 542–547 (1975)
32. Stamatatos, E.: A survey of modern authorship attribution methods. J. Am. Soc. Inf. Sci. Technol. **60**(3), 538–556 (2009)
33. Sun, J., Yang, Z., Liu, S., Wang, P.: Applying stylometric analysis techniques to counter anonymity in cyberspace. J. Netw. **7**(2), 259–266 (2012)
34. Yang, Y., Pedersen, J.O.: A comparative study on feature selection in text categorization. In: Proceedings of Fourteenth International Conference on Machine Learning, vol. 97, pp. 412–420, Nashville, TN, USA (1997)
35. Zheng, R., Li, J., Chen, H., Huang, Z.: A framework for authorship identification of online messages: writing-style features and classification techniques. J. Am. Soc. Inf. Sci. Technol. **57**(3), 378–393 (2006)
36. Zheng, R., Qin, Y., Huang, Z., Chen, H.: Authorship analysis in cybercrime investigation. In: Chen, H., Miranda, R., Zeng, D.D., Demchak, C.C., Schroeder, J., Madhusudan, T. (eds.) ISI 2003. LNCS, vol. 2665, pp. 59–73. Springer, Heidelberg (2003)

Multilevel Syntactic Parsing Based on Recursive Restricted Boltzmann Machines and Learning to Rank

Jungang Xu[✉], Hong Chen, Shilong Zhou, and Ben He

University of Chinese Academy of Sciences, Beijing, China
{xujg,benhe}@ucas.ac.cn,
{chenhong113,zhoushilong12}@mails.ucas.ac.cn

Abstract. Syntactic parsing is one of the central tasks in Natural Language Processing. In this paper, a multilevel syntactic parsing algorithm is proposed, which is a three-level model with innovative combinations of existing mature tools and algorithms. First, coarse-grained syntax trees are generated with general algorithms, such as Cocke-Younger-Kasami (CYK) algorithm based on Probabilistic Context Free Grammar (PCFG). Second, Recursive Restricted Boltzmann Machines (RRBM) are constructed, which aim at extracting feature vector through training syntax trees with deep learning methods. At last, Learning to Rank (LTR) model is trained to get the most satisfactory syntax tree and furthermore turn the parsing problem into a typical retrieval problem. Experiment results show that our method has achieved the state-of-the-art performance on syntactic parsing task.

Keywords: Deep learning · Recursive restricted boltzmann machines · Learning to rank · Multilevel syntactic parsing

1 Introduction

Recently, learning algorithms for deep structures, such as Deep Belief Networks (DBN) [1–3] and Deep Boltzmann Machines (DBMs) [4] have been proposed and successfully applied to various machine learning tasks such as image processing [5, 6], speech recognition [7, 8] and natural language processing (NLP) [9]. The applications in NLP mainly focus on two aspects. Firstly, recent studies have reaped huge fruits through the training of distributed representations, which can better capture the semantic similarities between words [10–12] and have been successfully applied to tackle all kinds of natural language processing tasks [13, 14]. Secondly, deep structure models have been constructed to make further improvement for the NLP tasks [15] increasingly.

Learning to Rank (LTR) is a supervised learning method for ranking, which has been widely applied to various areas, such as information retrieval (IR), recommender system, and machine translation. Conventional models, such as the Vector Space Model (VSM) [16], Boolean Model [17] and Probability Model [18], are typically not good at combining different dimensions of features, and LTR was designed for better tackling this issue. Moreover, LTR has a series of matured and profound theoretical support and shows more concern on how to design a reasonable and tractable model, with its

© Springer International Publishing Switzerland 2016
M. Chau et al. (Eds.): PAISI 2016, LNCS 9650, pp. 53–62, 2016.
DOI: 10.1007/978-3-319-31863-9_4

parameters trained by the iterative optimization. LTR can be divided into three categories based on the organization of the training data: Pointwise, Pairwise and Listwise [19].

As one of the crucial task of NLP, syntactic parsing is responsible for the mediation between linguistic structure and sentence meaning. Moreover, the results of syntactic parsing are indispensable for other NLP tasks such as relation extraction, semantic role labeling [20] and paraphrase detection [13]. Syntax description is a series of independent terms such as VP for verb phrases or PP for prepositional phrases etc. Recent studies have poured much attention into fine-grained syntactic categories for better representation of the phrases in the domain of both manual feature engineering [21] and automatic learning [22]. However, these methods infer little representation of phrase meaning and semantic similarity if anything. To go further, lexicalized parsers [23, 24], which associate each category with a lexical item, are proposed to give a fine-grained notions of semantic similarity. This approach has to use complex shrinkage estimation schemes to deal with the sparsity of observations of the lexicalized categories.

In this paper, we will introduce a novel algorithm for syntactic parsing and have observed significant improvements in accuracy with much lower computational cost. The details of our proposed approach are described in Sect. 2. In Sect. 3, we show the experimental results of our approach on the PTB corpus. Finally, we conclude the paper and discuss the directions of the future work.

2   Multilevel Syntactic Parsing

The main task of syntactic parsing is to generate the best syntax tree according to the existing syntactic rules. In 2000, Collins [25] proposed a re-ranking method, which firstly uses an existing model to generate several most possible parsing trees of a sentence to be analyzed, and then exploits a second model to re-rank the possibilities of the trees furthermore, and lastly picks the optimal parsing tree as the final result. Similarly, we take further exploration by deep learning techniques and build a multilevel model to select the most satisfying syntax trees.

2.1   Level 1: The Coarse-Grained Selection of Syntax Tree

In the first level, we make the coarse-grained selection to obtain k candidate syntax trees. Our model firstly exploits the CYK algorithm, which is based on probabilistic context-free grammar (PCFG), in which each rule has its own probability. It assumes that each context-free rule is independent and thus the probability of a sentence can be calculated directly by multiplying the probability of each rule. And then we can eliminate structural ambiguity of the sentence through selecting the most likely parsing tree. The probability of PCFG can be obtained by either a well-analytical corpus or raw corpus directly. Typically, we can use the Inside-Outside algorithm as the partial solution for grammar ambiguity. However, the algorithm is slow and local maxima of a problem are much more. Young et al. [26] suggest that many more nonterminals are required than theoretically necessary to get good grammar learning, which is of much less efficiency. As the standard algorithm for the probabilistic parsing, the probabilistic CYK algorithm is essentially a bottom-up parsing, using breadth-first search strategy as well as a dynamic programming algorithm that can be carried out in parallel without backtrack and redundant operations. When it is applied to any probabilistic context-free grammar, which should be firstly transformed into a weak equivalence of the Chomsky normal form grammar, the traditional sequencing of grammar rules actually acts as a speed bump. Fortunately, by rearranging rules of a given traditional grammar and establishing an index table, searching steps are sharply decreased.

2.2 Level 2: Feature Extraction of Candidate Syntactic Trees Based on RRBM Model

In the second level, we extract the low-dimension features of candidate syntactic trees for the further analysis, and construct a feature extraction model based on Recursive Restrict Boltzmann Machines (RRBMs).

As shown in Fig. 1, each parent-child pair, associated with one syntactic rule, constructs one RBM structure. In other words, different RBM parameters denote different syntactic rules. For each node of a RBM, we assign a vector and concatenate the child ones as the visible layer and assign embeddings from [11] to leaves in the syntactic tree directly. Consequently, the vector of the top node can be obtained by training all the RBMs in a bottom-up and layer-wise way based on the tree structure, which contains the information of semantic as well as the stable syntactic structure. In detail, the input layer of $P^{(1)} \rightarrow BC$ can be concatenated in order from vector b to vector c. Afterwards, the parameter matrix $W^{(B \wedge P^{(1)}, C \wedge P^{(1)})} = \left[W^{(B \wedge P^{(1)})} W^{(C \wedge P^{(1)})} \right]$ will be constructed, where $B \wedge P^{(1)}$ denotes the child node B of the parent node $P^{(1)}$. And then the vector $p^{(1)}$ of the node $P^{(1)}$ can be obtained by the Contrastive Divergence (CD) algorithm [3]. Similarly, the vector $p^{(2)}$ of the node $P^{(2)}$ (the top one in Fig. 1) can also be obtained in the same way with conservation of the grammatical information [10, 15].

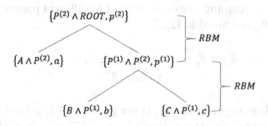

Fig. 1. The structure of RRBM

The construction of the RRBM model is based on the hypothesis that each syntax tree has its own energy, and lower energy means more stable distribution. To train the RRBM model, we firstly construct the energy function for RBM in Eq. 1.

$$E(v, h|\theta) = -\sum_{i=1}^{n} a_i v_i - \sum_{j=1}^{m} b_j h_j - \sum_{i=1}^{n} \sum_{j=1}^{m} v_i W_{ij} h_j \qquad (1)$$

Where $\theta = \{W_{ij}, a_i, b_j\}$ are real parameters, and W_{ij} denotes the weight between the visible unit i and the hidden unit j, and a_i denotes the bias of the visible unit while b_j is the bias of the hidden unit.

We exploit the CD algorithm for training parameters of the RBM. At beginning of the CD algorithm, we initialize the visible units with input samples and then figure out all the binary state of hidden units. Afterwards, we can obtain the probabilities of $v_i = 1$ and reconstruct the visible layer. The update rules for training a RBM can be denoted in Eq. 2, 3 and 4.

$$\Delta W_{ij} = \epsilon \left(\langle v_i h_j \rangle_{data} - \langle v_i h_j \rangle_{recon} \right) \tag{2}$$

$$\Delta a_i = \epsilon \left(\langle v_i \rangle_{data} - \langle v_i \rangle_{recon} \right) \tag{3}$$

$$\Delta b_j = \epsilon \left(\langle h_j \rangle_{data} - \langle h_j \rangle_{recon} \right) \tag{4}$$

Where ϵ denotes the learning rate and $\langle \cdot \rangle_{recon}$ indicates the model distribution of one-step reconstruction. We call this recursive training process as Recursive Pre-training (RP) and then we can construct the energy function for the syntax tree in Eq. 5.

$$E_T(v, h; \theta) = - \sum_{i \in R} h_i v_i \tag{5}$$

Where R denotes the set of RBM in the syntax tree, and its gradient with respect to θ is defined in Eq. 6.

$$\Delta \theta = \frac{\partial E_T(v, h; \theta)}{\partial \theta} \tag{6}$$

To minimize the objective function, we use the diagonal variant of AdaGrad [27], which exploits diverse learning rates with respect to different parameters. At time step t, the ith parameter $\theta_{t,i}$ is defined in Eq. 7.

$$\theta_{t,i} = \theta_{t-1,i} - \frac{\alpha}{\sqrt{\sum_{\tau=1}^{t} g_{\tau,i}^2}} g_{t,i} \tag{7}$$

Where α is the learning rate, and $g_{t,i}$ is the gradient of $\theta_{t,i}$. Then we can refine the model according to the gradient function Eqs. 6 and 7, and then obtain the training parameters, which are several parameter matrices of syntactic rules. Each right part of the grammar rules corresponds to a parameter matrix, which resembles the method exploited in CVG model [15]. Accordingly, for a given syntactic tree, the abstract features of the top layer can be extracted from the parameter matrices mentioned above.

After training the RRBM model, the vector of the root represents the steady-state feature of the current tree. Afterwards, we use the model to extract features, which are low-dimension vectors, including the semantic information and the structure information of the syntax tree.

2.3   Level 3: Re-ranking with LTR

In the third level, we convert the issue of the fine-grained selection of syntax trees to a retrieval issue with the LTR algorithm in detail.

The Query Mapping of Parsing. Before ranking with the LTR model, we preprocess and map all the syntax trees into the LTR training set. The training data set is composed of sentences and the corresponding standard trees, from which we firstly take the former k trees in level 1, and then obtain the training parameters for the RRBM model in level 2, lastly we make the fine-grained selection through the candidate syntax trees in level 3. Note that we can introduce any feature despite of how these features will be mapped and derived in the completed syntax trees. For instance, the top vector of the syntax tree is just the feature extracted from our model. Each syntax tree in level 1 can be mapped to a document among the query results, and its feature vector can be gained by the RRBM model in level 2. Lastly the query results can be ranked by LTR algorithm in level 3.

The Correlation Tag of Syntax Trees. LTR algorithm is a supervised learning method, so when we use LTR, the data should be marked to construct the ranking issue. Furthermore, we take F1 measure as the evaluation criteria for the syntax trees, and the higher value of F1 represents the higher rank of the syntax tree. In our model, we use 5 levels to mark the ranking results, take F1 that is greater than 95 as 4, take F1 that is between 90 and 95 as 3, etc. Then we can generate labeled LTR data set for training or testing LTR models.

Analysis on LTR. For syntactic parsing, we concern whether the filtered trees are correct as well as whether the trees we retrieve and standard trees are in a reasonable distance. We assume that syntax trees with higher scores come out at the top of the list. LTR algorithms are divided into Pointwise, Pairwise and Listwise, where the Pointwise method converts the ranking problem to the classification and regression for documents, which can lead to drastic decline in the ranking performance, and it is not suitable for our model for lack of the ranking information. While Pairwise and Listwise take the order among documents into account and treat the whole query as one instance to train, and it has the promising improvements on the quality of the first document.

3   Experiments

3.1   Corpus

The corpus exploited in our experiment is Penn Treebank (PTB), which is a text collection parsed and labeled with syntax and grammar structure manually. The linear sequence of words included in one sentence infers a hierarchical structure which is represented with a tree in general. Quantities of sentences and their corresponding trees constitute the Treebank, and we take a subset of PTB for our experiment, where 3,895 syntax trees are included.

3.2   Evaluation Criteria

The evaluation criteria of syntactic structure plays an important role in syntactic analysis and we introduce the widely used measures in PARSEVAL system, which are defined in Eqs. 8 and 9, into our experiment.

$$P = \frac{n(\hat{y})}{n(y)} \tag{8}$$

$$R = \frac{n(\hat{y})}{n(n_t)} \tag{9}$$

Where P denotes the precision measure and R denotes the recall measure, and $n(\hat{y})$ is the number of phrases that the model has correctly recognized while $n(y)$ denotes the total number of phrase that the model has recognized and $n(n_t)$ indicates the whole phrases in the corpus. To make better balance between the precision measure and the recall measure, we adopt the F1 measure, which is defined in Eq. 10.

$$F1 = \frac{2PR}{P + R} \tag{10}$$

Note that phrases that are correctly recognized should be not only with the right boundary, but also with accurate labels. Therefore, we make displacement of P and R with the Labeled Precision (LP) and the Labeled Recall (LR) [28].

3.3   Design of Experiments

We compare our method with the original PCFG and CVG algorithm on the accuracy rate and computational cost and also make tests on several combinations with Pairwise algorithm, such as RankNet, RankBoost [29], ListNet, LambdaMART [30] and AdaRank [31], for the LTR algorithm. Initially, we select 100 candidate trees by the CYK algorithm based on the PCFG algorithm in level 1. Then we mark these candidate trees with 5-level labels as described in Sect. 2.3. During the training process of RRBM in level 2, 3,382 trees are used as the training set, generating 3,382 queries and each one has 100 labeled results, which are also the input data of the LTR model for training and testing in level 3. To make a comprehensive analysis of the performance of the algorithm, the training set adopted in our model is divided into the long- and short-sentence training sets, which are bounded by 25 words. Moreover, the long-sentence training set includes 1,489 syntax trees for training and 206 syntax trees for testing while the short-sentence training set includes 1,893 trees for training and 307 trees for testing. In consideration of the fact that the PCFG algorithm does not generate a model, we compare it with the CVG algorithm. Meanwhile, all experiments are performed on a server with an Intel Quad Core CPU of 2.83 GHz, 4.0 GB Memory, and 1 TB Hard Disk.

3.4   Results and Analysis

Firstly, we make experiments with the full data. As shown in Table 1, deep learning methods have made some improvements compared with PCFG. The combination of RRBM and Listwise, including the combination of RRBM and ListNet, the combination of RRBM and LambdaMART as well as the combination of RRBM and AdaRank, is less impressive than the CVG algorithm. However, the combination of RRBM and Pairwise, including the combination of RRBM and RankBoost, and the combination of RRB and RankNet performs better than the CVG algorithm. We obtain that Pairwise algorithms have better performance on selecting satisfying syntax trees based on features extracted by the RRBM model. Secondly we conduct experiments with the long and short sentences. As shown in Tables 2 and 3, the results are consistent with the performance on the full data, the combination of RRBM and RankBoost can achieve favorable performance while other deep learning methods combined with LTR algorithm may be not stable under certain circumstances. Consequently, we can summarize that RRBM

Table 1. Performance on full data set (in percentage)

Algorithm	Precision	Recall	F1
PCFG	85.73	84.91	85.32
CVG	87.78	88.15	87.96
RRBM + ListNet	86.34	85.81	85.93
RRBM + LambdaMART	86.80	86.49	86.51
RRBM + AdaRank	86.73	85.95	86.21
RRBM + RankBoost	**92.62**	**93.22**	**92.21**
RRBM + RankNet	**94.86**	**95.18**	**94.95**

Table 2. Performance on the long sentences set (in percentage)

Algorithm	Precision	Recall	F1
PCFG	82.69	82.35	82.52
CVG	85.73	85.90	85.82
RRBM + ListNet	81.12	81.42	81.19
RRBM + LambdaMART	83.45	83.27	83.28
RRBM + AdaRank	80.93	80.94	80.86
RRBM + RankBoost	**87.15**	**87.79**	**87.40**
RRBM + RankNet	82.73	83.03	82.80

Table 3. Performance on the short sentences set (in percentage)

Algorithm	Precision	Recall	F1
PCFG	89.72	88.23	88.97
CVG	87.31	87.49	87.40
RRBM + ListNet	88.72	87.57	87. 96
RRBM + LambdaMART	88.29	87.90	87.95
RRBM + AdaRank	89.35	88.07	88.54
RRBM + RankBoost	**90.66**	**89.77**	**90.05**
RRBM + RankNet	89.81	88.76	89.12

can generate vectors containing the information of structures in syntax trees, with which we can train a robust model to select the correct syntax tree.

We also compare deep learning models with the CVG algorithm on training time cost as shown in Fig. 2, and it is obvious that our models consume much less time than that CVG algorithm does. Digging deeper into our method, we can find the reason that the parameters we need for the fine-tuning process are far fewer and we have pretrained these parameters by RRBM, which can save vast computation time and avoid being stuck in the local optimum caused by BP as well.

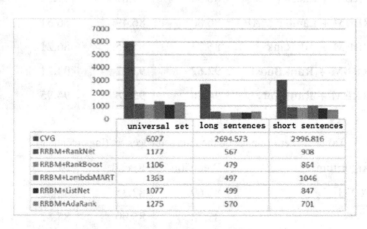

	universal set	long sentences	short sentences
■ CVG	6027	2694.573	2996.816
■ RRBM+RankNet	1177	567	908
■ RRBM+RankBoost	1106	479	864
■ RRBM+LambdaMART	1363	497	1046
■ RRBM+ListNet	1077	499	847
■ RRBM+AdaRank	1275	570	701

Fig. 2. Training time comparison of the algorithms (in second)

4   Conclusion

In this paper, a novel multilevel parsing algorithm is proposed which can generate k candidate syntax trees through coarse-grained selection in the first place and then re-rank for the selection of the most satisfying syntax tree. By layer-wise construction of the model, each RBM can be denoted with energy and completes the unsupervised training through building and optimizing the energy function. For the whole syntax tree, we build the RRBM model to conduct bottom-up pre-training of RBMs recursively and make up-bottom back propagation to refine RRBM by minimizing the energy function. Afterwards, we obtain candidate vectors, which contain the information of syntactic structure and semantic similarities. At last, we exploit LTR algorithms to make selection among the candidate trees. Experimental results show that our method can achieve favorable performance. Currently, PCFG algorithm is directly exploited to make the coarse-grained selection. In future works, we will do a deeper exploration among improved PCFG algorithms to further alleviate pressures on later training process and enhance the quality of the coarse-grained selection.

Acknowledgements. This work is supported in part by the National Natural Science Foundation of China under Grant no. 61372171.

References

1. Mnih, A., Hinton, G.: Three new graphical models for statistical language modelling. In: Proceedings of the 24th International Conference on Machine Learning, pp. 641–648. ACM (2007)
2. Hinton, G.E., Salakhutdinov, R.R.: Reducing the dimensionality of data with neural networks. Science **313**(5786), 504–507 (2006)
3. Hinton, G.E., Osindero, S., The, Y.W.: A fast learning algorithm for deep belief nets. Neural Comput. **18**(7), 1527–1554 (2006)
4. Salakhutdinov, R., Hinton, G.E.: Deep boltzmann machines. In: International Conference on Artificial Intelligence and Statistics, pp. 448–455 (2009)
5. Krizhevsky, A., Hinton, G.E.: Learning multiple layers of features from tiny images. Computer Science Department, University of Toronto, Technical report 1 (4), 7 (2009)
6. Krizhevsky, A., Hinton, G.E.: Using very deep autoencoders for content-based image retrieval. In: ESANN. Citeseer (2011)
7. Mohamed, A., Dahl, G.E., Hinton, G.: Acoustic modeling using deep belief net-works. IEEE Trans. Audio Speech Lang. Process. **20**(1), 14–22 (2012)
8. Hinton, G.E., Deng, L., Yu, D., Dahl, G.E., Mohamed, A., Jaitly, N., Senior, A., Vanhoucke, V., Nguyen, P., Sainath, T.N., et al.: Deep neural networks for acoustic modeling in speech recognition: the shared views of four research groups. Sig. Process. Mag. IEEE **29**(6), 82–97 (2012)
9. Salakhutdinov, R., Hinton, G.E.: Semantic hashing. Int. J. Approximate Reasoning **50**(7), 969–978 (2009)
10. Bengio, Y., Ducharme, R., Vincent, P., et al.: A neural probabilistic language model. J. Mach. Learn. Res. **2003**(3), 1137–1155 (2003)
11. Mikolov, T., Sutskever, I., Chen, K., Corrado, G.S., Dean, J.: Distributed representa-tions of words and phrases and their compositionality. In: Advances in Neural Information Processing Systems, pp. 3111–3119 (2013)
12. Hinton, G.E.: Learning distributed representations of concepts. In: Proceedings of the Eighth Annual Conference of the Cognitive Science Society, pp. 1–12 (1986)

13. Collobert, R., Weston, J.: A unified architecture for natural language processing: Deep neural networks with multitask learning. In: Proceedings of the 25th International Conference on Machine Learning, pp. 160–167. ACM (2008)
14. Huang, E.H., Socher, R., Manning, C.D., et al.: Improving word representations via global con-text and multiple word prototypes. In: Proceedings of the 50th Annual Meeting of the Association for Computational Linguistics: Long Papers, vol. 1, pp. 873–882. Association for Computational Linguistics (2012)
15. Socher, R., Bauer, J., Manning, C.D., Ng, A.Y.: Parsing with compositional vector grammars. In: Proceedings of the ACL Conference (2013)
16. Salton, G., Wong, A., Yang, C.S.: A vector space model for automatic indexing. Commun. ACM 18(11), 613–620 (1975)
17. Manning, C.D., Raghavan, P., Schütze, H.: Introduction to Information Retrieval. Cambridge University Press, Cambridge (2008)
18. Zhai, C.X.: Statistical language models for information retrieval. Synth. Lect. Hum. Lang. Technol. 1(1), 1–141 (2008)
19. Xia, F., Liu, T.Y., Wang, J., et al.: Listwise approach to learning to rank: theory and algorithm. In: Proceedings of the 25th International Conference on Machine Learning, pp. 1192–1199. ACM (2008)
20. Gildea, D., Palmer, M.: The necessity of parsing for predicate argument recognition. In: Proceedings of the 40th Annual Meeting on Association for Computational Linguistics, ACL 2002, pp. 239–246. Association for Computational Linguistics, Stroudsburg (2002)
21. Klein, D., Manning C.D.: Accurate unlexicalized parsing. In: Proceedings of the 41st Meeting of the Association for Computational Linguistics, pp. 423–430 (2003)
22. Petrov, S., Barrett, L., Thibaux, R., Klein, D.: Learning accurate, compact, and interpretable tree annotation. In: Proceedings of the 21st International Conference on Computational Linguistics and the 44th Annual Meeting of the Association for Computational Linguistics, pp. 433–440. Association for Computational Linguistics (2006)
23. Charniak, E.: A maximum-entropy-inspired parser. In: Proceedings of the 1st North American chapter of the Association for Computational Linguistics Conference, pp. 132–139. Association for Computational Linguistics (2000)
24. Collins, M.: Head-driven statistical models for natural language parsing. Comput. Linguist. 29(4), 589–637 (2003)
25. Collins, M., Koo, T.: Discriminative reranking for natural language parsing. Comput. Linguist. 31(1), 25–70 (2005)
26. Younger, D.H.: Recognition and parsing of context-free languages in time n 3. Inform. Control 10(2), 189 (1967)
27. Duchi, J., Hazan, E., Singer, Y.: Adaptive subgradient methods for online learning and stochastic optimization. J. Mach. Learn. Res. 12, 2121–2159 (2011)
28. Abney, S., Flickenger, S., Gdaniec, C., et al.: Procedure for quantitatively comparing the syntac-tic coverage of English grammars. In: Proceedings of the Workshop on Speech and Natural Language, pp. 306–311. Association for Computational Linguistics (1991)
29. Freund, Y., Iyer, R., Schapire, R.E., et al.: An efficient boosting algorithm for combining preferences. J. Mach. Learn. Res. 2003(4), 933–969 (2003)
30. Burges, C.J.C.: From ranknet to lambdarank to lambdamart: an overview. Learning 2010(11), 23–581 (2010)
31. Xu, J., Li, H.: AdaRank: a boosting algorithm for information retrieval. In: Proceedings of the 30rd Annual International ACM SIGIR Conference on Research and Development in Information Retrieval, pp. 391–398. ACM (2007)

Stratified Over-Sampling Bagging Method for Random Forests on Imbalanced Data

He Zhao[1]([✉]), Xiaojun Chen[2], Tung Nguyen[3], Joshua Zhexue Huang[2],
Graham Williams[4], and Hui Chen[1]

[1] Shenzhen Institutes of Advanced Technology, Chinese Academy of Sciences,
Shenzhen, China
{he.zhao,hui.chen1}@siat.ac.cn
[2] College of Computer Science and Software Engineering, Shenzhen University,
Shenzhen, China
{xjchen,zx.huang}@szu.edu.cn
[3] Faculty of Computer Science and Engineering, Thuyloi University,
Hanoi, Vietnam
tungnt@tlu.edu.vn
[4] Australian National University, Canberra, Australia
Graham.Williams@togaware.com

Abstract. Imbalanced data presents a big challenge to random forests
(RF). Over-sampling is a commonly used sampling method for imbalanced data, which increases the number of instances of minority class
to balance the class distribution. However, such method often produces
sample data sets that are highly correlated if we only sample more minority class instances, thus reducing the generalizability of RF. To solve
this problem, we propose a stratified over-sampling (SOB) method to
generate both balanced and diverse training data sets for RF. We first
cluster the training data set multiple times to produce multiple clustering results. The small individual clusters are grouped according to their
entropies. Then we sample a set of training data sets from the groups
of clusters using stratified sampling method. Finally, these training data
sets are used to train RF. The data sets sampled with SOB are guaranteed to be balanced and diverse, which improves the performance of
RF on imbalanced data. We have conducted a series of experiments, and
the experimental results have shown that the proposed method is more
effective than some existing sampling methods.

Keywords: Imbalanced data · Classification · Stratified sampling ·
Random forests

1 Introduction

In recent years, class-imbalanced data become very common in real-world applications, such as rare disease prediction, fraud detection and risk management.
The class is imbalanced when a data set has a large number of instances belonging to one class whereas the numbers of instances in other classes are small.

© Springer International Publishing Switzerland 2016
M. Chau et al. (Eds.): PAISI 2016, LNCS 9650, pp. 63–72, 2016.
DOI: 10.1007/978-3-319-31863-9_5

In imbalanced data sets, the class with the largest number of instances is denoted as the majority class, and other small classes are called the minority classes. The machine learning methods face problems in building classifiers from the class-imbalanced data. The prediction performance of the classifier is often low, because conventional machine learning methods tend to build models biased to majority class. Hence, classifiers do not accurately predict the instances belonging to the minority class [8]. There are two main approaches to deal with imbalanced data: (i) machine learning methods [12,18] and (ii) the training data resampling methods [5,15].

Random forests (RF) [3] is an ensemble machine learning method composed of un-pruned decision trees. RF is widely used in data mining domain and has achieved good performance when dealing with both regression and classification problems [1]. RF also works well to build the model for high-dimensional data [13,16,17]. However, this model often results in low prediction performance when applied to imbalanced data, because it uses the bagging method [2] to generate sub-data sets for building trees. A bagged sub-data set consists of instances randomly drawn with equal probability from the input data. This bagged set contains fewer or even none of instances from the minority classes. The tree constructed from such sub-data set increases prediction error when applied to minority class cases. As a result, the prediction accuracy of the RF model is reduced.

The sampling method has been an important research topic in learning the model to classify the imbalanced data [7,8,10]. Rather than over-sampling the minority class with replacement, Chawla et al. [5] proposed SMOTE to combine down-sampling and over-sampling with synthetic minority class to boost the performance on imbalanced data. Afterwards, a boosting version of SMOTE (SMOTE-Boost) [6] was proposed to further improve the prediction performance. Chen et al. [7] proposed the balanced RF model (BRF) and weighted RF model (WRF) to learn imbalanced data. The idea of BRF is to combine the under-sampling technique and ensemble method. For each repetition, the same number of instances from the minority class and the majority class are randomly drawn with replacement, respectively. WRF follows the idea of cost sensitive learning, where a heavier penalty is placed on the node-splitting condition and terminal node class weights for misclassifying the minority class. Their BRF and WRF models achieved the better prediction performance when compared with some other one-sided sampling method such as SMOTE. Krawczyk et al. [11] proposed a cost-sensitive ensemble method (EG2Ensemble) for improving minority class prediction. They used EG2 [14] to build cost-sensitive trees, and used random subspaces approach [9] and a genetic algorithm to form the ensemble.

In this paper, we propose a stratified over-sampling (SOB) method to generate a set of balanced training data sets which are lowly correlated. We first cluster the training data set multiple times to produce multiple clustering results. The small individual clusters are grouped according to their entropies. Then we sample a set of training data sets from the groups of clusters using stratified sampling method. Finally, these training data sets are used to train RF. The sampling probability in each group is proportional to the entropy of the cluster and inversely proportional to the size of the class. Therefore, the data set sampled with SOB is balanced and diverse.

The rest of this paper is organised as follows. Section 2 gives a brief introduction to RF. Section 3 presents the stratified over-sampling bagging method to generate a set of training data sets for RF on imbalanced data. The experimental results are reported and discussed in Sect. 4. Conclusions and future work are given in Sect. 5.

2   The Random Forest Model

The original random forest (RF) model [3] is to build an ensemble (i.e., a collection) of decision tree models. When combining decision trees into an ensemble as a classification model the class assigned to a new instance is the class that the majority of trees assign to the instance.

For a multi-class problem with q class labels $\mathcal{L} = \{l_1, l_2, \ldots, l_q\}$, denote the training data set as $D = \{d_i\}_{i=1}^n = \{(x_i, y_i)\}_{i=1}^n$, where x_i is an m-dimensional vector and $y_i \in \mathcal{L}$. The key development of RF is the random selection of both instances and features throughout the tree building process:

1. First, t subsets, $S = \{S_1, S_2, \ldots, S_t\}$, are obtained from random instances with the replacement of the original data set D, which is called bagging [2];
2. Then a decision tree h_j is built with a decision tree algorithm (e.g., CART [4]) on each set S_j, such that the RF model is $h = \{h_j\}_{j=1}^t$. Instead of using all m features as candidates when selecting the best split for any node during tree building, a subspace $m' \ll m$ is chosen at random. Besides, there is no pruning in the process of tree building;
3. Finally, the mode of the classes from the t trees is taken as the class label for the new instance x, that is $h(x) = Mo(\{h_j(x)\}_{j=1}^t)$.

3   Stratified Over-Sampling Bagging Method

In this section, we propose the Stratified Over-sampling Bagging (SOB) method. We introduce an over-sampling bagging method first to produce balanced sampling data from imbalanced data. We further improve this process by proposing the stratified over-sampling bagging method to generate sample data sets that are both balanced and diverse. Finally, we analyze the sampling probability in the stratified over-sampling bagging method.

3.1   Over-Sampling Bagging for Imbalanced Data

In bagging [2], each instance is equally sampled with replacement, i.e., for each instance $d \in D$, the probability of its being sampled is $p(d) = 1/n$. Such method faces a big problem for imbalanced data. To deal with this problem, we propose an over-sampling bagging method which increases the number of minority class instances to balance the class distribution. We first count the frequency of each class $l \in \mathcal{L}$ as $Count(l)$, and then we compute the sampling probability of an instance d as

$$p(d) \propto \frac{1}{Count(l)}, \tag{1}$$

where the class label of d is l, $p(d)$ is normalized such that

$$\sum_{d \in D} p(d) = 1. \tag{2}$$

$p(d)$ in (1) is inversely proportional to the size of the class it belongs to. The instances of minority classes will be assigned big probabilities while instances of majority classes will be assigned small probabilities. In bagging, instead of equally sampling with replacement, we sample each instance with probability $p(d)$. We can verify that the expected probability of each class in the sample result is equal. Therefore, the above over-sampling bagging method can produce training data with expected balanced class distribution.

However, since the minority classes only takes a small proportion of the all classes, the training data sets produced by the above method may have a high correlation, thus reducing the generalizability of RF built on them. To deal with this problem, we further propose a Stratified Over-sampling Bagging (SOB) method in the next subsection.

3.2 The Stratified Over-Sampling Bagging (SOB)

Let the $n \times m$ matrix D be the training data set with n instances and m features. Given the number of produced clusterings r and the number of sampling data sets t, we propose a stratified over-sampling method to generate t informative and balanced training data sets. As shown in Fig. 1, the new method consists of three main steps:

1. **Clustering** the training data set multiple times to produce multiple clustering results;
2. **Grouping** individual small clusters into a number of groups according to their entropies;
3. **Sampling** t training data sets from the groups of clusters using the stratified sampling method.

Finally, the t training data sets will be used to train the RF model.

In the first step, we produce r clustering results from the original data set D. Let $\mathcal{C}$ be the set of the resulting clusters, initialized as $\emptyset$. For each of the l-th clustering result, we first randomly select the number of clusters $k \in [q, \min\{4q, n/10\}]$. We then run the k-means method to produce k disjoint clusters $\mathcal{P} = \{p_1, ..., p_k\}$ and update $\mathcal{C} = \mathcal{C} \bigcup \mathcal{P}$.

In the second step, for each cluster $c \in \mathcal{C}$, we first measure the class distribution in the clustering by computing its entropy as

$$Entropy(c) = -\sum_{l \in \mathcal{L}} p(l|c) \log p(l|c). \tag{3}$$

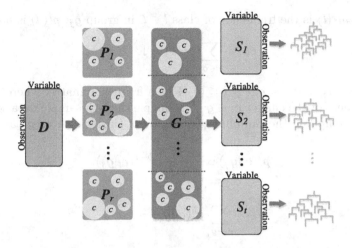

Fig. 1. The illustration of the stratified over-sampling bagging method

The higher the entropy, the evener the classes in the cluster are distributed. We then partition C into q disjoint groups $G = \{g_i\}_{i=1}^q$, such that

$$\begin{cases} Entropy(c_i) > Entropy(c_j), \forall c_i \in g_{i'}, c_j \in g_{j'}, i' > j' \\ |g_i| \equiv |g_j|, \forall i, j \in [1, q] \end{cases}, \tag{4}$$

where $|g_i|$ is the number of clusters in g_i.

In the third step, we propose a stratified over-sampling bagging method to sequentially sample t training data sets. For the first data set, we randomly sample n instances from G with replacement. For each sample, we first randomly select one group g from G with equal probability. Then we sample a cluster c_l from g_j with the following probability

$$p^1(c|g_j) = \frac{Entropy(c)}{\sum_{c \in g} Entropy(c)}. \tag{5}$$

To produce balanced training data set, we prefer to select the cluster with high entropy values, because the class distribution in a cluster with high entropy will be more balanced than that in a cluster with low entropy.

Finally we sample an instance d from c with the following probability

$$p(d|c) = \frac{p^j(d)}{\sum_{d \in c} p^j(d)}, \tag{6}$$

where $c \in g_j$ and $p^j(d)$ is the sampling probability defined in one group, which is computed as

$$p^j(d) \propto \frac{1}{Count^j(l)}, \tag{7}$$

where $Count^j(l)$ is the frequency of class $l \in \mathcal{L}$ in group g_j, $p^j(d)$ is normalized such that

$$\sum_{d \in g_j} p^j(d) = 1. \tag{8}$$

Let $F = \{f_{jl} | 1 \leq j \leq q, 1 \leq l \leq |g_j|\}$ be a frequency matrix where f_{jl} is the sampling times of the cluster $c_l \in g_j$. To sample the b-th sample data set, where $1 < b \leq t$, we update the sampling probability of cluster c_l from g_j as

$$p^b(c_l|g_j) \propto \frac{1}{1 + f_{jl}} p^{b-1}(c_l|g_j), \tag{9}$$

such that

$$\sum_{c_l \in g_j} p^b(c_l|g_j) = 1. \tag{10}$$

Algorithm 1. Stratified over-sampling bagging

Input:

The data set D, the number of clusterings r, the number of impurity groups q and the number of sample data sets t.

Output: t sample data sets.

1: //Clustering
2: $\mathcal{C} = \emptyset$.
3: **for** $i = 1$ to r **do**
4:     Randomly select $k \in [q, \min\{4q, n/10\}]$;
5:     Partition D into k clusters $\mathcal{P} = \{p_1, ..., p_k\}$;
6:     Update $\mathcal{C} = \mathcal{C} \bigcup \mathcal{P}$.
7: **end for**
8: //Grouping
9: **for** each $c \in \mathcal{C}$ **do**
10:     Compute $Entropy(c)$ according to (3).
11: **end for**
12: Partition $\mathcal{C}$ into q disjoint groups $G = \{g_i\}_{i=1}^q$ according to (4).
13: //Sampling
14: Initialize each element in F as zero.
15: **for** $z = 1$ to t **do**
16:     $S_z = \emptyset$.
17:     //Randomly sampling n samples with replacement
18:     **for** $i = 1$ to n **do**
19:         Randomly sample a group $g \in G$ with equal probability;
20:         Randomly sample a cluster $c \in g_j$ with $p^b(c|g)$ according to (5);
21:         Randomly sample an instance $d \in c$ with $p(d|c)$ according to (6);
22:         Update F and $S_z = S_z \bigcup d$.
23:     **end for**
24:     Update $\mathcal{S} \Leftarrow \mathcal{S} \cup \{S_z\}$.
25: **end for**
26: **return** $\mathcal{S}$.

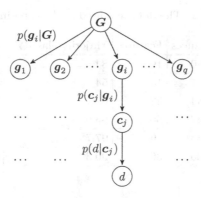

Fig. 2. The sampling probability in SOB

We can see from (9) that if a cluster c_l is already sampled too many times during the sampling for the previous $b-1$ data sets, the sampling probability of c_l will be decreased, thus increasing the diverse of the sample data sets. Then the b-th sample data set can be sampled with the same method used in the first data set, except that we sample a cluster c_l from g_j with $p^b(c_l)$. The new method is formulated in Algorithm 1.

3.3   Analysis of the Sampling Probability of SOB

As shown in Fig. 2, in the SOB method, an instance is sampled with three steps: (1) sample a group g from G with equal probability, i.e., $p(g|)G = 1/q$; (2) sample a cluster c from g with $p^b(c|G)$; and finally (3) sample an instance d from c with probability $p(d|c)$. For the first sample data, the sampling probability for each instance d in each g is

$$p^1(d|g) = p(c|G)p(d|c) \propto \frac{Entropy(c)}{Count^j(l)}. \tag{11}$$

From (11), we can see that the sampling probability in the SOB method is proportional to the entropy of the cluster the instance is assigned to, and inversely proportional to the size of the class it belongs to. Therefore, the data set sampled with SOB is balanced and informative. But the number of clusters with high entropies are often small. If we only sample data from clusters with high entropies, the correlation of resulting t sample data sets can be high. In SOB, we sample each group with equal probability so that the t sample data sets are diverse.

4   Experiments

4.1   Experimental Settings and Data Sets

We selected six real-life data sets [7,11], as listed in Table 1, from the UC Irvine Machine Learning Repository[1]. We can see that all six data sets are imbalanced.

---

[1] http://archive.ics.uci.edu/ml/index.html.

Table 1. The data sets used in the experiments

Dataset	Instances	Features	Majority class (%)	Minority class (%)
Hepatitis	80	19	84	16
Heart disease	297	13	54	46
Mushroom	5644	22	62	38
Oil	937	50	95.6	4.4
Mammography	11183	6	97.7	2.3
Satimage	6435	36	90.3	9.7

The minority class in Mammography takes only 2.3%, which is hard to build an accurate classification model. The number of instances in Mammography is more than eleven thousand.

In this section, we compared SOB with four methods (i.e., EG2Ensemble [11], SMOTE-Boost [6], BRF and WRF [7]) on the six data sets. We used the k-means method to generate 200 clusters ($r = 200$) for SOB on all data sets ($t = 300$). On each data set, we produced 300 sample data sets. The Breiman's random forests method [3] was used to build the classification model.

4.2   Experimental Results

Table 2 lists the comparison results of SOB, EG2, BoostEG2 and EG2Ensemble on three data sets. We can see that SOB achieved the best result on almost all cases, except for recall measure on Hepatitis. SOB even obtained 100% accuracy on Mushroom.

The results of Oil, Mammography and Satimage data sets are presented in Tables 3, 4 and 5. From the results on Oil and Satimage, we can see that all the results from SOB are greater than the others except for the recall measure. Though the results of SOB on Mammography are not the best, they are

Table 2. Experimental results on three data sets. Numbers in bold are the best results.

	Hepatitis	Heart disease	Mushroom	
SOB	50.50	**78.35**	**100**	Recall
	92.59	**81.30**	**100**	Specificity
EG2	47.35	68.50	73.50	Recall
	68.35	74.45	89.10	Specificity
BoostEG2	55.00	70.20	80.00	Recall
	73.25	80.80	92.30	Specificity
EG2Ensemble	**55.50**	73.20	80.34	Recall
	73.10	81.20	92.60	Specificity

Table 3. Experimental results on Oil. Numbers in bold are the best results.

	Recall	Specificity	Precision	F_1-measure
SOB	52.45	**98.16**	**54.96**	**52.32**
BRF	73.2	91.6	28.57	41.10
WRF	**92.7**	82.4	19.39	32.07

Table 4. The results on Mammography. Numbers in bold are the best results.

	Recall	Specificity	Precision	F_1-measure
SOB	64.79	**99.62**	73.27	68.65
BRF	**76.54**	98.21	50.51	60.86
WRF	65.38	99.57	**78.34**	**71.28**

Table 5. The results on Satimage. Numbers in bold are the best results.

	Recall	Specificity	Precision	F_1-measure
SOB	71.88	**97.30**	**73.98**	**72.91**
SMOTE-Boost	67.87	97.25	72.68	70.19
BRF	77.00	93.56	56.31	65.05
WRF	**77.48**	94.56	60.55	67.98

comparable with BRF and WRF. In a summary, the stratified over-sampling bagging method improves the performance of random forests on imbalanced data.

5   Conclusions and Future Work

In this paper, we have presented a new stratified over-sampling bagging method to generate the set of balanced and diverse training data sets, which are used to build the RF model on imbalanced data. In the new method, we first produce multiple clustering results and group the clusters in these findings into a set of groups according to their entropies. Then we propose to use a stratified sampling method to sequentially sample a group, and then a cluster from the group, and finally an instance from the cluster, with respect to sample probabilities. We have conducted experiments on six imbalanced data sets, experimental results have shown the improvement of our method in prediction accuracy.

In the future work, we plan to further improve the performance of the proposed method on classifying imbalanced data. Other techniques, such as subspace weighting methods, can be introduced into this process. Moreover, we will test our method on more applications.

Acknowledgments. This work was supported by Guangdong Fund under Grant No. 2013B091300019, NSFC under Grant No. 61305059 and No. 61473194, and Natural Science Foundation of SZU (Grant No. 201432).

References

1. Banfield, R.E., Hall, L.O., Bowyer, K.W., Kegelmeyer, W.P.: A comparison of decision tree ensemble creation techniques. IEEE Trans. Pattern Anal. Mach. Intell. **29**(1), 173–180 (2007)
2. Breiman, L.: Bagging predictors. Mach. Learn. **24**(2), 123–140 (1996)
3. Breiman, L.: Random forests. Mach. Learn. **45**(1), 5–32 (2001)
4. Breiman, L., Friedman, J.H., Olshen, R.A., Stone, C.J.: Classification and Regression Trees. CRC Press, Boca Raton (1984)
5. Chawla, N.V., Bowyer, K.W., Hall, L.O., Kegelmeyer, W.P.: SMOTE: synthetic minority over-sampling technique. J. Artif. Intell. Res. **16**, 321–357 (2002)
6. Chawla, N.V., Lazarevic, A., Hall, L.O., Bowyer, K.W.: SMOTEBoost: improving prediction of the minority class in boosting. In: Lavrač, N., Gamberger, D., Todorovski, L., Blockeel, H. (eds.) PKDD 2003. LNCS (LNAI), vol. 2838, pp. 107–119. Springer, Heidelberg (2003)
7. Chen, C., Liaw, A., Breiman, L.: Using random forest to learn imbalanced data. Technical report TR.666, University of California, Berkeley, California (2004)
8. He, H., Garcia, E.A.: Learning from imbalanced data. IEEE Trans. Knowl. Data Eng. **21**(9), 1263–1284 (2009)
9. Ho, T.K.: The random subspace method for constructing decision forests. IEEE Trans. Pattern Anal. Mach. Intell. **20**(8), 832–844 (1998)
10. Jo, T., Japkowicz, N.: Class imbalances versus small disjuncts. SIGKDD Explor. Newsl. **6**(1), 40–49 (2004)
11. Krawczyk, B., Wozniak, M., Schaefer, G.: Improving minority class prediction using cost-sensitive ensembles. In: 16th Online World Conference on Soft Computing in Industrial Applications (2011)
12. Liu, Y., Yu, X., Huang, J.X., An, A.: Combining integrated sampling with SVM ensembles for learning from imbalanced datasets. Inf. Process. Manag. **47**(4), 617–631 (2011)
13. Nguyen, T., Huang, J.Z., Nguyen, T.T.: Two-level quantile regression forests for bias correction in range prediction. Mach. Learn. **101**(1–3), 325–343 (2015)
14. Núñez, M.: The use of background knowledge in decision tree induction. Mach. Learn. **6**, 231–250 (1991)
15. Seiffert, C., Khoshgoftaar, T.M., Hulse, J.V.: Hybrid sampling for imbalanced data. In: Proceedings of the IEEE International Conference on Information Reuse and Integration 2008, Las Vegas, Nevada, USA, pp. 202–207, 13–15 July 2008
16. Xu, B., Huang, J.Z., Williams, G.J., Wang, Q., Ye, Y.: Classifying very high-dimensional data with random forests built from small subspaces. Int. J. Data Warehous. Min. **8**(2), 44–63 (2012)
17. Ye, Y., Wu, Q., Huang, J.Z., Ng, M.K., Li, X.: Stratified sampling for feature subspace selection in random forests for high dimensional data. Pattern Recogn. **46**(3), 769–787 (2013)
18. Yen, S.J., Lee, Y.S.: Cluster-based under-sampling approaches for imbalanced data distributions. Expert Syst. Appl. **36**(3), 5718–5727 (2009)

Revisiting Attribute Independence Assumption in Probabilistic Unsupervised Anomaly Detection

Sunil Aryal[1,2]([✉]), Kai Ming Ting[2], and Gholamreza Haffari[1]

[1] Clayton School of Information Technology, Monash University,
Victoria, Australia
{sunil.aryal,gholamreza.haffari}@monash.edu
[2] School of Engineering and Information Technology,
Federation University, Victoria, Australia
{sunil.aryal,kaiming.ting}@federation.edu.au

Abstract. In this paper, we revisit the simple probabilistic approach of unsupervised anomaly detection by estimating multivariate probability as a product of univariate probabilities, assuming attributes are generated independently. We show that this simple traditional approach performs competitively to or better than five state-of-the-art unsupervised anomaly detection methods across a wide range of data sets from categorical, numeric or mixed domains. It is arguably the fastest anomaly detector. It is one order of magnitude faster than the fastest state-of-the-art method in high dimensional data sets.

Keywords: Fast anomaly detection · Independence assumption · Big data

1 Introduction

Let database D be a collection of n data instances $(\mathbf{x}_1, \mathbf{x}_2, \cdots, \mathbf{x}_n)$. Each instance $\mathbf{x}$ is a m-dimensional vector $\langle x_1, x_2, \cdots, x_m \rangle$ where each component is either a numeric attribute $x_i \in \mathcal{R}$ ($\mathcal{R}$ is a real domain) or a categorical attribute $x_i \in \{v_{i_1}, \cdots, v_{i_w}\}$ (where v_{i_j} is a label out of w possible labels for attribute x_i). The problem of anomaly detection is to identify anomalous instances which are significantly different from the majority of instances in the database.

In the literature, anomaly detection has two main approaches [1]: (i) the supervised approach classifies an instance in either anomaly or normal class by using a classification model trained on a labelled training set; (ii) the unsupervised approach trains an anomaly detector from unlabelled training data, and identifies anomalies based on their anomaly scores. In many real-world applications, labelled training data are difficult to obtain; and thus anomalies have to be identified using an unsupervised approach. In this paper, we focus on the unsupervised approach to anomaly detection.

© Springer International Publishing Switzerland 2016
M. Chau et al. (Eds.): PAISI 2016, LNCS 9650, pp. 73–86, 2016.
DOI: 10.1007/978-3-319-31863-9_6

As databases are currently growing rapidly in terms of volume and dimensionality, identifying anomalous patterns in massive databases is a challenging task. Traditional distance or density based unsupervised methods using nearest neighbours such as kNN [2] and LOF [3] are not applicable in large databases because of their high time complexity in the order of $O(n^2m)$. Recently, simpler and more efficient methods such as iForest [4] and the nearest neighbour in a small subsamples (Sp) [5] are proposed for numeric domains. In categorical domains, anomaly detection methods have been largely based on frequent patterns (e.g., FPOF [6] and COMPREX [7]) which do not scale up to high dimensionality and large data size.

In probabilistic approach, instances in low density region are considered as anomalies, i.e., anomalies have a low probability to be generated from the distribution of normal instances. The simplest efficient probabilistic approach of estimating multivariate probability as the product of univariate probabilities has been used for anomaly detection in some domains [1,8]. The intuition behind this Simple univariate Probabilistic Anomaly Detector (we call it SPAD) is that an anomalous instance is significantly different from normal instances in a few attributes where it has low probability.

Despite its simplicity, effectiveness and efficiency, SPAD is not considered as a benchmark to compare the performance of recently proposed efficient unsupervised anomaly detectors [4–7]. They are shown to run orders of magnitude faster than traditional kNN based methods and produce better or competitive detection accuracy. But, it is not clear if they are more effective and efficient than the simplest traditional probabilistic method SPAD.

In this paper, we show that SPAD performs competitively to or better than all the state-of-the-art anomaly detection methods mentioned above in a wide range of 25 data sets from categorical only, numeric only, and mixed domains. It runs one order of magnitude faster than the simplest nearest neighbour anomaly detector Sp [5] in data sets with high dimensionality. In categorical domains, it runs up to five orders of magnitude faster than the existing state-of-the-art categorical based methods [6,7].

The rest of the paper is organised as follows. SPAD and five widely used state-of-the-art anomaly detection methods are discussed in Sect. 2, followed by experimental evaluations in Sect. 3. The parametrized version of SPAD is described in Sect. 4, and the conclusions are provided in the last section.

2   Related Work

In this section, we review six unsupervised anomaly detectors including five widely used state-of-the-art anomaly detection methods and SPAD. The first three are designed primarily for numeric domains and the last three are mainly for categorical domains. The pertinent details of these methods are described in the following six subsections.

2.1 Local Outlier Factor (LOF)

Breunig et al. (2000) [3] proposed a method based on relative density of an instance with respect to its k-neighbourhood. Let $N^k(\mathbf{x})$ be the set of k-nearest neighbours of $\mathbf{x}$, $d(\mathbf{x}, \mathbf{y})$ be the distance between $\mathbf{x}$ and $\mathbf{y}$ and $d^k(\mathbf{x}, D)$ is the distance between $\mathbf{x}$ and its kth-NN in D. The anomaly score of $\mathbf{x}$ is defined as follows:

$$s_{lof}(\mathbf{x}) = \frac{\sum_{\mathbf{y} \in N^k(\mathbf{x})} lrd(\mathbf{y})}{|N^k(\mathbf{x})| \times lrd(\mathbf{x})} \tag{1}$$

where $lrd(\mathbf{x}) = \frac{|N^k(\mathbf{x})|}{\sum_{\mathbf{y} \in N^k(\mathbf{x})} \max(d^k(\mathbf{y}, D), d(\mathbf{x}, \mathbf{y}))}$.

It requires a distance measure to compute all pairwise distances between instances in D. Euclidean distance $\left(d_{euc}(\mathbf{x}, \mathbf{y}) = \sqrt{\sum_{a=1}^{m}(x_a - y_a)^2} \right)$ is the most widely used distance measure. In the case of categorical attribute a, $x_a - y_a = 0$ if $x_a = y_a$; and 1 otherwise. An alternative measure in categorical domains advocated by Boriah et al. [9] is Occurrence Frequency (OF): $d_{of}(\mathbf{x}, \mathbf{y}) = \frac{1}{m} \sum_{a=1}^{m}(x_a - y_a)$ where $x_a - y_a = 0$ if $x_a = y_a$; and $x_a - y_a = \log \frac{n}{f(x_a)} \times \log \frac{n}{f(y_a)}$ otherwise; where $f(x_a)$ is the frequency of the categorical label x_a in D.

2.2 Isolation Forest (iForest)

Instead of density, iForest [4] employs an isolation mechanism to isolate every instance in the given training set. This is done efficiently by random axis-parallel partitioning of the data space in a tree structure until every instance is isolated. A set of t trees is constructed, each tree T_i is built using a subsample randomly selected from D. The anomaly score of an instance $\mathbf{x}$ is measured as the average path length over t trees as follows:

$$s_{iforest}(\mathbf{x}) = \frac{1}{t} \sum_{i=1}^{t} \ell_i(\mathbf{x}) \tag{2}$$

where $\ell_i(\mathbf{x})$ is the path length of $\mathbf{x}$ in tree T_i.

The intuition is that anomalies are more susceptible to isolation. Isolation using trees yields that anomalies have shorter average path lengths than normal instances. iForest is designed for numeric domains only. In this paper, we show that it can be effective in categorical and mixed domains as well, by simply converting each categorical label into a binary $\{0, 1\}$ attributes and treating them as numeric attributes (see the empirical evaluation in Sect. 3).

2.3 Sampling (Sp)

Instead of searching k-nearest neighbour in D, Sugiyama and Borgwardt (2013) [5] proposed to search the nearest neighbour (i.e., $k = 1$) in a small random subsample $\mathcal{D} \subset D$ to detect anomalies. The anomaly score is the distance to the nearest neighbour in $\mathcal{D}$, defined as follows:

$$s_{sp}(\mathbf{x}) = \min_{\mathbf{y} \in \mathcal{D}} d(\mathbf{x}, \mathbf{y}) \tag{3}$$

They have shown that the method called Sp with a very small subsample ($\psi = 20$) performs better than or competitive to LOF; but it runs several orders of magnitude faster. In [5], Sp is used only in numeric domains with the euclidean distance. In this paper, we evaluated Sp in categorical and mixed domains as well with the euclidean and Occurrence Frequency (OF) distance measures [9].

2.4   Frequent Pattern Outlier Factor (FPOF)

He et al. (2005) [6] proposed an anomaly detection method for categorical domains based on frequent patterns. It uses Apriori algorithm [10] to generate frequent itemsets of maximum size η with minimum support threshold δ in D, denoted as $FPS(D, \eta, \delta)$. The score of an instance $\mathbf{x}$ is estimated as follows:

$$s_{fpof}(\mathbf{x}) = \frac{\displaystyle\sum_{z \subseteq \mathbf{x} \ \wedge \ z \in FPS(D,\eta,\delta)} support(z)}{|FPS(D, \eta, \delta)|} \tag{4}$$

where z is a frequent itemset, $z \subseteq \mathbf{x}$ denotes that z is contained in $\mathbf{x}$, and $support(z)$ is the support of z.

The intuition of Eq. 4 is that $\mathbf{x}$ is more likely to be an anomaly if it has a few or none of the frequent itemsets, i.e., the lower the score, the more likely $\mathbf{x}$ is an anomaly.

2.5   Pattern Based Compression Technique (COMPREX)

Recently, Akoglu et al. (2012) proposed a pattern-based compression technique called COMPREX for anomaly detection in categorical data [7]. It builds a collection of dictionaries (code tables) $CT_1, CT_2, \cdots, CT_k$ which are learnt from data using disjoint subsets of highly correlated features based on information gain. The anomaly score of $\mathbf{x}$ is defined as the cost of encoding $\mathbf{x}$ using the code tables.

$$s_{cmprx}(\mathbf{x}) = \sum_{F \in \mathcal{P}} \sum_{p \subseteq \pi_F(\mathbf{x})} L(code(p)|CT_F) \tag{5}$$

where $\mathcal{P} = \{F_1, F_2, \cdots, F_k\}$ is a set of disjoint partitions of the feature set, p is a pattern, $\pi_F(\mathbf{x})$ is a projection of $\mathbf{x}$ into feature subset F, $code(p)$ is a code word corresponding to p and $L(\cdot)$ is the length of a code word.

The intuition is that the higher the cost of encoding $\mathbf{x}$, the more likely it is to be an anomaly. Even though it produces better detection accuracy than other categorical methods, it is limited to low dimensional data sets because of its high time complexity.

2.6   Simple Probabilistic Method (SPAD)

In probabilistic approach, instance $\mathbf{x}$ is an anomaly if the probability of $\mathbf{x}$, $P(\mathbf{x})$, is low. An estimate of $P(\mathbf{x})$ requires a large amount of data (even in a moderate number of dimensions) which is usually infeasible in many applications. Assuming the attributes are independent of each other (e.g. naive Bayes [11]), it can be decomposed as:

$$\hat{P}(\mathbf{x}) = \prod_{i=1}^{m} \hat{P}(x_i) \tag{6}$$

The one-dimensional $\hat{P}(x_i)$ can be estimated from D using the Laplace-corrected estimate as: $\hat{P}(x_i) = \frac{f(x_i)+1}{n+w_i}$, where $f(x_i)$ is the occurrence frequency of x_i in D, and w_i is the number of possible values of x_i. The same estimation can be used in numeric domains by converting numeric attributes into categorical attributes through discretisation[1].

In order to avoid floating point overflow, instances are ranked using logarithm of $P(\mathbf{x})$ in Eq. 6, effectively using the summation of the logarithm of univariate probabilities $P(x_i)$ as:

$$s_{spad}(\mathbf{x}) = \sum_{i=1}^{m} \log \hat{P}(x_i) \tag{7}$$

Instances are ranked in ascending order based on their probabilities — instances having low probability, which are ranked at the top, are likely to be anomalies. Note that SPAD is parameter-free, like COMPREX. Goldstein and Dengel (2012) used a similar score as defined in Eq. 7, though they have called it a histogram-based method and evaluated it using three numeric data sets only [8].

Compared with the commonly used frequent pattern based approach, SPAD shares two common features: they are all based on categorical domains and employ probability or frequency as the basis for detecting anomalies. Yet, SPAD is significantly simpler and requires no search; whereas the frequent pattern approach requires an expensive search.

SPAD is scalable to both high dimension and large data sets as it just needs to store the frequency count of every categorical label which can be done in a single pass of data in $O(nm)$ time and requires $O(mw)$ space (where w is the average number of labels in each dimension). Having constructed the frequency count table (in a preprocessing step), calculating the score of an instance just needs a table look-up which costs $O(m)$. Hence, the total runtime complexity of ranking n instances is $O(nm)$ which is cheaper than iForest (if $m < t \log_2 \psi$ which is generally the case unless the dimensionality of the data is very high in the order of thousands); and it is ψ times faster than Sp.

The time and space complexities of the above six anomaly detectors are presented in Table 1.

---

[1] Though $\hat{P}(\cdot)$ can be estimated directly in numeric domains, it is a lot easier and faster to do it in categorical domains.

Table 1. Time and space complexities

Anomaly detector	Time complexity	Space complexity
LOF	$O(n^2 m)$	$O(nm)$
iForest	$O(nt \log_2 \psi)$	$O(t\psi)$
Sp	$O(nm\psi)$	$O(m\psi)$
FPOF	$O(n2^m)$	$O(2^m)$
COMPREX	$O(nm^2)$	$O(m^2)$
SPAD	$O(nm)$	$O(mw)$

w: The average number of categorical labels in each dimension.

3   Empirical Evaluation

In this section, we present the evaluation results of the six anomaly detection methods discussed in Sect. 2: LOF-L2 (LOF using the euclidean distance), LOF-OF (LOF using the occurrence frequency based distance), iForest, Sp-L2 (Sp with the euclidean distance), Sp-OF (Sp with the occurrence frequency based distance), FPOF, COMPREX and SPAD.

We used 25 benchmark data sets from UCI machine learning data repository [12] with categorical only, numeric only and mixed attributes. The data sets were from different domains ranging from health and medicine, text, email filtering to digit recognition. Many of the data sets used were from classification problems. We converted them to anomaly detection problems by considering some larger classes (1 or more) as normal and some smaller classes (1 or more) as anomalies. In the case where classes were more or less uniformly distributed, anomalies were random samples from classes which are not used as normal. They have different data characteristics in terms of data size ($366 \leq n \leq 5$ million), dimensionality ($5 \leq m \leq 4670$) and anomaly proportion ($0.5\% \leq p \leq 35\%$). The characteristics of data sets used are given in Table 2.

Some preprocessing is required for some algorithms. For iForest which can handle numeric attributes only, each categorical label was converted into a binary attribute [13] (where 0 represents the absence of the label and 1 represents the presence); and all converted binary attributes are treated as numeric. For algorithms that can handle categorical attributes only, numeric attributes were converted into categorical using a modified version of equal width bins which divides the range $[\mu_j - 3\sigma_j, \mu_j + 3\sigma_j]$ into b equal width bins (where μ_j and σ_j are the mean and standard deviation of data values in dimension j). This version is used to reduce the distortion due to outliers. This discretisation technique with $b = 5$ was used to discretise numeric attributes for all categorical methods: SPAD, LOF-OF, Sp-OF, FPOF and COMPREX.

Parameters in all the algorithms were set to the suggested values in respective papers. For LOF, k was set to a commonly used value of 10 [3,5]. The parameter ψ in Sp was set to 20 as suggested in [5]. Parameters η and δ in FPOF were set

Table 2. Characteristics of data sets. n: data size, m: # attributes, m_{num}: # numeric attributes, m_{cat}: # categorical attributes

Name	n	m	m_{num}	m_{cat}	anomaly %
Chess	4580	6	0	6	0.5
Nursery	4648	8	0	8	7
Solar	1066	9	0	9	1
Mushroom	4429	22	0	22	5
Dermatology	366	33	0	33	5.5
Reuters	4297	4134	0	4134	9
Newsgroup	5668	4670	0	4670	14
Advertise	3279	1558	3	1555	14
Arrhythmia	452	279	206	73	14.5
Kddcup99	64759	41	34	7	6.5
U2r	60821	41	34	7	0.5
Census	299285	40	7	33	6
Hypothyroid	3772	29	7	22	7.5
Sick	3772	27	6	21	6
Annthyroid	7200	21	6	15	7.5
Lymph	148	18	3	15	4
Covertype	287128	12	10	2	1
Linkage	5749132	7	2	5	0.5
Breastw	683	9	9	0	35
Spambase	2964	57	57	0	6
Mnist	20444	96	96	0	3.5
Har	7032	561	561	0	20
Secom	1567	590	590	0	6
Isolete	730	617	617	0	1.5
Mfeat	410	649	649	0	2.5

to 5 and 0.1, respectively, as suggested in [6]. Parameters $t = 100$ and $\psi = 256$ were used in iForest, as suggested in [4].

All the algorithms were implemented in JAVA using the WEKA [13] platform, except COMPREX for which we used the MATLAB implementation provided by the authors [7][2]. All the experiments were conducted in a Linux machine with 2.27 GHz processor and 8 GB memory.

We used area under the receiver operating curve (AUC) as the measure of anomaly detection performance. AUC = 1 if all the anomalies are at the top of the ranked list; and a randomly ranked list will yield AUC = 0.5. We conducted

---

[2] http://www3.cs.stonybrook.edu/~leman/pubs.html.

10 runs for each of the randomised methods: iForest, Sp-L2 and Sp-OF; and reported the average AUC (and time). A significance test was conducted using confidence interval based on two standard errors over 10 runs. A win or loss between two algorithms is counted only if the difference is significant; otherwise it is a draw.

The AUC of all methods in the 25 data sets is provided in Table 3. Note that FPOF and COMPREX did not complete in data sets with high number of dimensions and/or large data size in 24 h.

The overall performance is summarised using the average rank, shown in the bottom three rows in Table 3. SPAD is the best method having an average rank of 1.8; and the closest contenders are COMPREX which has a rank of 2.3 (based on the result of 15 data sets only), and iForest which has a rank of 2.9.

In the last row of Table 3, the pairwise win:loss:draw (w:l:d) counts of contenders against SPAD clearly shows that LOF-L2, LOF-OF, FPOF, iForest, Sp-L2 and Sp-OF produced significantly worse AUC than SPAD (i.e., they have more losses than wins). COMPREX produced competitive detection accuracy in comparison with SPAD, where they have the same number of wins and losses on 15 data sets in which COMPREX completed.

SPAD produced the best (or equivalent to the best) AUC in 13 data sets followed by iForest (7), COMPREX (5), Sp-OF (3), Sp-L2 (2), FPOF (2), LOF-L2 (1) and LOF-OF (0). Note that SPAD is one of the top three performers in almost every data set we have used, shown in the second last row in Table 3. The only exception is Mnist, where SPAD is the fourth best performer. The two closest contenders are iForest and COMPREX which are one of the top three performers in 17 out of 25 data sets and 11 out of 15 data sets, respectively.

The total runtime, including preprocessing (discretisation or nominal to binary conversion), model building (if required), ranking n instances and computing AUC, is provided in Table 4. Note that the direct comparison of the runtime of COMPREX with other methods is not fair as it was implemented in MATLAB and others were implemented in JAVA. It is included in the table to provide an idea about its runtime.

SPAD was faster than all the contenders in all data sets, except in a few small data sets where it ran slight slower than Sp. Note that SPAD ran one order of magnitude faster than its closest contender Sp and iForest in the two highest dimensional data sets (Reuters and Newsgroup) which have more than 4000 attributes; and SPAD ran one to five orders of magnitude faster than LOF, FPOF and COMPREX in all data sets. The only exception is Lymph, compared with LOF-L2. Note that the runtime of LOF-L2 presented here are without using indexing scheme to expedite the k nearest neighbour search [2,14,15]. We did not use indexing because they are not effective as the number of dimensions increases and they do not work with non-metric distance such as OF.

It is interesting to note the longest time a method took in all these data sets: SPAD took less than 20 s to complete in Linkage, the largest data set having more than 5 million instances. Sp and iForest took about 50 s and 256 s, respectively, in the same data set. Ignoring the data sets in which they could not complete

Table 3. Anomaly detection performance (AUC). The average rank, the number of data sets on which a method is among the top three performers, and win:loss:draw (w:l:d) counts of a method against SPAD are included in the last three rows.

Data set	LOF-L2	LOF-OF	FPOF	CMPRX	iForest	Sp-L2	Sp-OF	SPAD
Chess	0.890	0.730	0.912	**0.995**	0.945	0.707	0.901	0.994
Nursery	0.446	0.365	**1.000**	**1.000**	**1.000**	0.759	0.480	**1.000**
Solar	0.588	0.651	0.979	0.968	0.943	0.801	0.855	***0.980**
Mushroom	0.371	0.709	0.922	**0.987**	0.907	0.892	0.892	0.936
Dermatology	**0.884**	0.418	0.641	0.810	0.726	0.638	0.397	0.727
Reuters	0.793	0.684	∘	•	0.861	0.865	0.802	***0.883**
Newsgroup	0.670	0.289	∘	•	0.684	0.708	0.635	***0.735**
Advertise	0.597	0.547	∘	•	0.689	**0.710**	0.673	**0.704**
Arrhythmia	0.729	0.798	∘	•	0.803	0.755	0.771	***0.813**
Kddcup99	0.553	0.804	0.997	0.943	**0.998**	0.925	0.920	0.996
U2r	0.626	0.799	•	0.968	0.986	0.985	0.943	***0.990**
Census	•	•	•	•	0.623	0.627	**0.724**	0.676
Hypothyroid	0.603	0.541	0.631	**0.688**	0.562	0.509	0.616	0.667
Sick	0.596	0.490	0.564	0.621	0.523	0.471	**0.641**	0.602
Annthyroid	0.669	0.599	0.688	0.696	0.632	0.503	0.603	***0.697**
Lymph	0.994	0.953	0.979	0.996	0.995	0.847	0.833	***0.997**
Covertype	0.520	0.479	0.838	•	**0.977**	0.930	0.925	**0.974**
Linkage	•	•	0.910	0.972	**1.000**	0.998	0.898	0.993
Breastw	0.435	0.822	**0.991**	0.990	0.987	0.952	0.971	0.990
Spambase	0.653	0.649	∘	**0.787**	0.785	0.658	0.717	0.770
Mnist	0.801	0.706	0.635	0.799	**0.835**	0.807	0.798	0.799
Har	0.318	0.998	∘	•	0.988	0.922	0.996	***1.000**
Secom	0.533	0.553	∘	•	0.537	0.534	0.548	***0.562**
Isolete	0.835	0.983	∘	•	**1.000**	**1.000**	**1.000**	**1.000**
Mfeat	0.755	0.722	∘	•	**0.947**	0.834	0.848	0.938
Avg. Rank	5.2	5.7	3.6	2.3	2.9	4.4	4.3	1.8
#Top3	3/23	3/23	7/14	11/15	17/25	10/25	7/25	24/25
w:l:d	2:21:0	0:23:0	2:11:1	6:6:3	5:16:4	2:21:2	2:21:2	–

Bold face: The best or equivalent to the best AUC.
∗: Significantly better detection performance of SPAD over all other contenders.
•: Did not complete in 24 h.
∘: Did not complete due to insufficient memory with 8 GB.

Table 4. Total runtime (in seconds).

Data set	LOF-L2	LOF-OF	FPOF	CMPRX	iForest	Sp-L2	Sp-OF	SPAD
Chess	5.14	14.69	0.49	148.61	0.59	0.12	0.24	0.13
Nursery	3.23	15.29	0.52	54.15	0.51	0.12	0.21	0.10
Solar	0.24	0.51	0.35	20.20	0.40	0.07	0.12	0.04
Mushroom	5.07	16.07	100.79	363.59	0.56	0.14	0.30	0.08
Dermatology	0.33	0.48	1848.42	128.62	0.66	0.05	0.06	0.03
Reuters	2198.30	1879.72	○	●	49.85	11.11	19.75	2.25
Newsgroup	5520.84	5126.25	○	●	76.33	31.68	43.98	5.26
Advertise	193.66	177.44	○	●	2.03	1.53	2.06	0.93
Arrhythmia	1.35	3.56	○	●	0.73	0.07	0.25	0.19
Kddcup99	3527.47	2419.10	79488.06	6272.76	4.47	1.10	1.91	0.86
U2r	1448.58	1987.84	●	4342.38	4.34	1.02	1.71	0.85
Census	●	●	●	●	25.84	5.60	13.29	3.40
Hypothyroid	9.57	15.93	1536.38	297.69	0.73	0.14	0.27	0.17
Sick	4.56	8.60	1076.37	149.14	0.66	0.14	0.26	0.15
Annthyroid	20.00	34.86	167.05	204.02	0.73	0.16	0.31	0.14
Lymph	0.08	0.24	1.79	28.03	0.21	0.02	0.03	0.02
Covertype	40586.59	85501.06	14.23	●	15.03	1.71	6.34	2.22
Linkage	●	●	137.87	43976.78	255.80	19.76	49.11	17.58
Breastw	0.54	0.72	0.37	28.45	0.39	0.03	0.07	0.09
Spambase	3.94	8.48	○	950.53	0.66	0.12	0.34	0.13
Mnist	366.48	2964.48	6072.35	13174.59	2.00	0.65	3.20	0.65
Har	398.22	1383.15	○	●	1.39	1.30	4.52	0.78
Secom	10.16	42.24	○	●	1.03	0.36	1.40	0.26
Isolete	4.61	22.90	○	●	1.25	0.18	0.90	0.14
Mfeat	1.75	7.33	○	●	1.33	0.11	0.55	0.12
Avg. Rank	5.0	5.8	5.9	7.0	4.1	1.6	2.8	1.4

Bold face: The best or equivalent to the best AUC.

∗: Significantly better detection performance of SPAD over all other contenders.

●: Did not complete in 24 h or 86400 s.

○: Did not complete due to insufficient memory with 8 GB.

in the experiments, LOF took the longest, close to one day, in Covertype which has less than 300000 instances; COMPREX took about half a day in Linkage; and FPOF took about 22 h in Kddcup99 which has 41 attributes and less than 65000 instances. Note also that COMPREX and FPOF would take more than a day to complete even for small data sets such as Isolete, Mfeat and Arrhythmia which have less than 1000 instances and 700 attributes.

4 Detecting Anomalies Which Rely on Multiple Attributes

Despite its strong assumption of attribute independence, SPAD produced superior performance than other state-of-the-art anomaly detectors in many data sets. But, it has a limitation in detecting anomalies which rely on multiple attributes as it does not capture the relationship between attributes. An example is shown in Fig. 1 where anomalies are not different from normal instances in any single attribute; but they exhibit outlierness only if both attributes are examined.

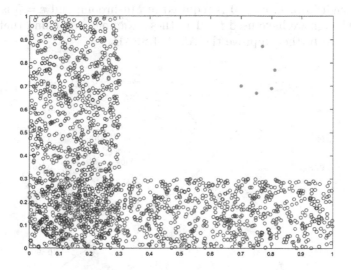

Fig. 1. An example where SPAD fails.

In order to handle such a situation, we propose a parametrized model called SPAD$_r$, in which data in the original m-dimensional space $\mathcal{X}$ are embedded into a new s-dimensional space $\mathcal{Z}$. Each dimension (or attribute) in $\mathcal{Z}$ is a product set of r dimensions (or attributes) in $\mathcal{X}$. The instances can be ranked using Eq. 7 in the new space $\mathcal{Z}$. In the new space, we are able to inspect data distribution along various combinations of the original attributes to identify those outliers which are otherwise difficult to identify.

To guarantee that all possible r attributes are represented in $\mathcal{Z}$, a set S of $\binom{m}{r}$ combinations is required, which is intractable even for moderate values of m and r. Instead, we randomly generate a subset of attribute combinations $\mathcal{S} \subset S$ ($|\mathcal{S}| = s \ll \binom{m}{r}$) as follows: A random order of m attributes (in $\mathcal{X}$) is generated. Then m sets of new attributes are formed from the ordered attributes through a moving window of size r (the attributes ordering is considered as a circular sequence). As an example, for $r = 2$ and a random ordering $[x_3, x_1, x_2]$ of three attributes in $\mathcal{X}$, three new attributes (z_1, z_2, z_3) are formed as $[(x_3 \wedge x_1), (x_1 \wedge x_2),$

$(x_2 \wedge x_3)]$. The above is repeated t times to produce $s = mt$ new attributes, where t is chosen such that $s \ll \binom{m}{r}$. Note that the simplest version with $r = 1$ and $t = 1$ is the SPAD used in Sect. 3.

SPAD_r has the advantage that the size of $\mathcal{S}$ is linear in t and m; and each attribute is used exactly r times. The time complexity for SPAD_r is $O(nmtr)$ and the space complexity is $O(mtw^r)$ since each attribute in $\mathcal{Z}$ has w^r labels.

If anomaly $\mathbf{x}$ relies on r attributes to be identified, and as long as one of the attributes in $\mathcal{Z}$ represents the r attributes in $\mathcal{X}$, then it can be detected by SPAD_r because it has low probability in that particular attribute in $\mathcal{Z}$.

In a data set, the appropriate setting of r in SPAD_r depends on the number of attributes required to detect anomalies. Figure 2 shows three examples where $r = 1$ is enough in Sick; $r = 10$ is required in Mushroom; and $r = 5$ is the best in Mfeat. Though we have used $t = 1$ in these examples, we found that in some cases, $t > 1$ can further improve the AUC of SPAD_r.

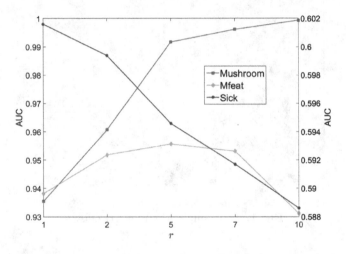

Fig. 2. AUC of SPAD_r w.r.t r ($t = 1$) (The curve of Sick is using the right y-axis) (Color figure online).

One can see some similarity of SPAD_r with subspace anomaly detection methods such as HICS [16]. The fundamental difference is that SPAD_r avoids the expensive search in HICS to find subspaces by considering random subspaces; and SPAD_r runs significantly faster than those complex subspace search based methods.

5 Conclusions

We show that the simple parameter-free anomaly detection method based on univariate probabilities is the fastest method and works as effective as, if not better than, five state-of-the-art anomaly detection methods. Its anomaly detection performance is consistent across different domains – categorical only, numeric only and mixed domains. It is the method of choice in big data and data streams, as far as we know.

The parametrized version can be used to further improve the detection performance of the parameter-free version in data sets where the detection of anomalies relies on multiple attributes.

Acknowledgments. We would like to thank Prof. Takashi Washio for providing very useful comments and suggestions. We are thankful to the anonymous reviewers for their critical comments to improve the quality of the paper.

References

1. Chandola, V., Banerjee, A., Kumar, V.: Anomaly detection: a survey. ACM Comput. Surv. **41**(3), 15: 1–15: 58 (2009)
2. Ramaswamy, S., Rastogi, R., Shim, K.: Efficient algorithms for mining outliers from large data sets. In: Proceedings of the 2000 ACM SIGMOD Conference on Management of Data, pp. 427–438 (2000)
3. Breunig, M.M., Kriegel, H.P., Ng, R.T., Sander, J.: LOF: identifying density-based local outliers. In: Proceedings of ACM SIGMOD Conference on Management of Data, pp. 93–104 (2000)
4. Liu, F., Ting, K.M., Zhou, Z.H.: Isolation forest. In: Proceedings of the Eighth IEEE International Conference on Data Mining, (ICDM), pp. 413–422 (2008)
5. Sugiyama, M., Borgwardt, K.M.: Rapid distance-based outlier detection via sampling. In: Proceedings of the 27th Annual Conference on Neural Information Processing Systems, Lake Tahoe, Nevada, United States, pp. 467–475 (2013)
6. He, Z., Xu, X., Huang, J.Z., Deng, S.: FP-outlier: frequent pattern based outlier detection. Comput. Sci. Inf. Syst. **2**(1), 103–118 (2005)
7. Akoglu, L., Tong, H., Vreeken, J., Faloutsos, C.: Fast and reliable anomaly detection in categorical data. In: Proceedings of the 21st ACM Conference on Information and Knowledge Management (CIKM), pp. 415–424 (2012)
8. Goldstein, M., Dengel, A.: Histogram-based outlier score (hbos): a fast unsupervised anomaly detection algorithm. In: Proceedings of the 35th German Conference on Artificial Intelligence (KI-2012), pp. 59–63 (2012)
9. Chandola, V., Boriah, S., Kumar, V.: Similarity measures for categorical data: a comparative study. Technical report TR 07–022, Department of Computer Science and Engineering, University of Minnesota, USA (2007)
10. Agrawal, R., Imieliński, T., Swami, A.: Mining association rules between sets of items in large databases. In: Proceedings of the 1993 ACM SIGMOD Conference on Management of Data, pp. 207–216 (1993)
11. Duda, R.O., Hart, P.E., Stork, D.G.: Pattern Classification, 2nd edn. Wiley-Interscience, New York (2000)
12. Bache, K., Lichman, M.: UCI machine learning repository, University of California, Irvine, School of Information and Computer Sciences (2013). http://archive.ics.uci.edu/ml
13. Hall, M., Frank, E., Holmes, G., Pfahringer, B., Reutemann, P., Witten, I.H.: The weka data mining software: an update. SIGKDD Explor. Newslett. **11**(1), 10–18 (2009)

14. Bay, S.D., Schwabacher, M.: Mining distance-based outliers in near linear time with randomization and a simple pruning rule. In: Proceedings of the Ninth ACM SIGKDD Conference on Knowledge Discovery and Data Mining, pp. 29–38 (2003)
15. Beygelzimer, A., Kakade, S., Langford, J.: Cover trees for nearest neighbor. In: Proceedings of the 23rd International Conference on Machine Learning, pp. 97–104 (2006)
16. Keller, F., Mller, E., Bhm, K.: HiCS: high contrast subspaces for density-based outlier ranking. In: Proceedings of ICDE, pp. 1037–1048. IEEE Computer Society (2012)

Incremental Privacy-Preserving Association Rule Mining Using Negative Border

Duc H. Tran[1,2(✉)], Wee Keong Ng[1], Y.D. Wong[2,3], and Vinh V. Thai[4]

[1] School of Computer Engineering, Nanyang Technological University,
Singapore, Singapore
ductran@ntu.edu.sg
[2] Maritime Institute, Nanyang Technological University, Singapore, Singapore
[3] School of Civil and Environmental Engineering,
Nanyang Technological University, Singapore, Singapore
[4] School of Business IT and Logistics, RMIT University, Melbourne, Australia

Abstract. Privacy preserving association rule mining can extract important rules from distributed data with limited privacy breaches. Protecting privacy in incremental maintenance for distributed association rule mining is necessary since data are frequently updated. In privacy preserving data mining, scanning all the distributed data is very costly. This paper proposes a new incremental protocol for privacy preserving association rule mining using negative border concept. The protocol scans old databases at most once, and therefore reducing the I/O time. We also conduct experiments to show the efficiency of our protocol over existing ones.

Keywords: Privacy preserving · Secure protocol · Association rule mining · Negative border · Incremental

1 Introduction

Since its inception, preserving privacy has become one of the major tasks of data mining area and has attracted tremendous interest among researchers. A variety of algorithms and techniques has been introduced to perform mining in secure manners. Traditionally, all these algorithms are designed with assumption that data is persistent. In case that there are updates, deletion of records, data mining process needs to run again. This certainly is impractical since mining on distributed data are so costly that it cannot be performed frequently.

Let see a scenario: n parties hold their horizontally partitioned data. They together perform a association rule mining as proposed by Kantarcioglu and Clifton [11]. After all parties have completed the algorithm and got results, another party with his own data wishes to join the task. The first n parties certainly want him to do the mining task for the accuracy of the mining results (the more data involved, the more accurate the result is). However, the first n parties have finished the mining task (often time-consuming) and are not really willing to start over again.

© Springer International Publishing Switzerland 2016
M. Chau et al. (Eds.): PAISI 2016, LNCS 9650, pp. 87–100, 2016.
DOI: 10.1007/978-3-319-31863-9_7

Another scenario is that, after a day or a week, all the parties collect more transaction data. They would like to update the mining result with less effort. The problem of updating association rules is first studied by Cheung with insertion operation in [4] and then updated with deletion and modification in [5]. However, the issue of maintaining association rules in privacy-preserving context is a challenge and to our best knowledge, there is no related work trying to solve this problem.

In this paper, we tackle this challenge using the original incremental techniques proposed by Cheung [4] and secure multi-party computation (SMC) in [8]. We propose a novel incremental association rule mining protocol that can be used when data of parties are updated or some new parties join the mining tasks. Our new protocol will update final results considering the old mining results and new data. In most cases, the protocol does not require to read the old data.

The rest of this paper is organized as follows. Section 2 demonstrates the problem. Section 3 reviews related work to our solution. Section 4 presents the background that is important to build our protocol. Section 5 introduces our novel protocol to incrementally perform association rule mining. We conduct experiments in Sect. 6. Finally, Sect. 7 gives the summary of this paper.

2   Problem Definition

In this section, we define the association rule mining problem. Here are some notations.

- $I = \{i_1..i_m\}$: a set of items
- DB: a set of transactions
- T: a transaction in DB. A transaction is a set of itemsets in I. We have $T \subseteq I$.
- L^{DB}: all frequent itemsets in DB.
- L_k^{DB}: all k-large itemsets in DB.

Given an itemset $X \subseteq I$. An association rule can be written as $X \Rightarrow Y$, where $X \subseteq I$, $Y \subseteq I$, and $X \bigcap Y = \emptyset$. The rule $X \Rightarrow Y$ is said to have support s if $s\%$ of transactions in DB contain $X \bigcup Y$. The rule holds is said to have confidence $c\%$ if $c\%$ transactions in DB that contain X also contains Y. The mining association rule problem is to discover all rules that have satisfied support and confidence thresholds.

Distributed Data. A database DB is horizontally partitioned in n sites $(S_1, ..., S_n)$. Their database are $DB_1, ..., DB_n$ respectively.

Assume that itemset X has count of $X.sup_i$ at each site S_i (i.e., $X.sup_i$ of the transactions contains X). Then we can compute global count of X as $X.sup = \sum_{i=1}^{n} X.sup_i$.

The goal of privacy preserving association rule mining is to discover aassociation rules satisfying thresholds, i.e., the sets L_k for all $k > 1$ without any privacy breach. A protocol is privacy preserving if no site should be able to learn extra information of any transaction at any other site other than the final results of mining tasks.

Incremental Mining. Assume that n parties have found the large itemsets for their data, presented in $\bigcup_{k=1}^{n} DB_k$. Now r new sites $S_{n+1}, S_{n+2}, ..., S_{n+r}$ want to do mining tasks with current n sites. The new sites have database DB_i, for sites $i = n + 1, ..., n + r$ respectively. It is simple that we can do this by asking all $n + r$ to do mining tasks again. However, this method takes long time and old sites may not be willing to run again. The purpose of incremental mining is to discover the new updated results without running algorithms all over old and new datasets. To protect the privacy of the old sites, the r new sites should not know the large itemsets L of the n old sites. On the other hand, the n old sites should also not know any other information of the r new sites except one stated as follows. However, due to the nature of this problem, privacy breach is unavoidable no matter how secure the protocol is. For instance, if an itemset X is large (small) in the n old sites but after updating it becomes small (large) in $n + r$ sites then all the old sites know that X is small (large) in the r new sites. Even if we run the protocol in [11] again, this conclusion is still correct.

Definition 1. *Let $X.sup_i$ be the support count of X in S_i. An itemset X is said to be globally large if $\sum_{k=1}^{n+r} X.sup_k \geq \sum_{k=1}^{n+r} DB_k \times s\%$. X is said to be group large in new sites if $\sum_{k=n+1}^{n+r} X.sup_k \geq \sum_{k=n+1}^{n+r} DB_k \times s\%$. X is said to be group large in old sites if $\sum_{k=1}^{n} X.sup_k \geq \sum_{k=1}^{n} DB_k \times s\%$.*

3 Related Works

3.1 Incremental Association Rule Mining

Association rule mining algorithms have been divided into two categories: Apriori-based and FP-tree-based. The problem of maintenance of association rules in large databases was first presented in 1996 by Cheung et al. with the FUP algorithm [4]. They then upgraded to the FUP2 algorithm [5] including not only addition but also deletion and modification of data. FUP2 is more efficient than FUP. In 1997, Thomas *et al.* [16] introduced a new method to boost the progress of the incremental mining using the negative border. Ayan *et al.* [2] presented a new method called UWEP (Update With Early Pruning), which exploits a dynamic look-ahead strategy. The list of large itemsets could be updated by checking itemsets if they are frequent in new database. In 2001, SWF method [12] was first demonstrated by Lee *et al.*. The method splits databases into partitions, and using a filtering threshold in each partition to create a list of candidate itemsets. Veloso *et al.* proposed a new method called ZigZag algorithm that makes use of *tidlist* and generates maximal frequent itemsets in new database to prevent from creating too many unnecessary candidates [18].

3.2 Privacy Preserving Association Rule Mining

While there has been a lot of related work in privacy-preserving data mining, due to space constraints, we only focus on the tightly related efforts. The method presented by Kantarcioglu *et al.* [11] is the first cryptography-based solutions for private distributed association rules mining, it assumes three or more parties, and they jointly do the distributed Apriori algorithm with the data encrypted. In the recent research papers [7,9,15,17,19], some privacy-preserving association rules schemes are proposed. These papers are similar and developed a secure multi-party protocol based on homomorphic encryption.

While the two above related issues have been well studied, there is a very limited work has been dedicated to simultaneously solve both incremental maintenance and privacy issue of association rule mining. Wong et al. proposed an incremental method to protect privacy for distributed association rule mining [19]. Their algorithm is based on the original FUP/FUP2 algorithm that may scan the old databases many times. In this paper, we present a new algorithm that requires the old database at most once. Our algorithm makes use of the negative border and cryptography techniques.

4 Preliminaries

4.1 Negative Border

In this subsection, we review the concept of *negative border* which was presented by Mannila and Toivonen [13].

Definition 2. *The negative border $Bd^-(L)$, of a collection of itemsets L is defined as follows: Given a collection $L \subseteq P(R)$ of sets, closed with respect to the set inclusion relation, the negative border $Bd^-(L)$ of L consists of the minimal itemsets $X \subseteq R$ not in L.*

Example 1. Let $R = \{A, B, ..., F\}$ and assume the collection L of frequent itemsets is L={{A},{B},{C},{F}, {A,B},{A,C}, {A,F},{C,F},{A,C,F}}. The negative border of the collection L contains the itemset $\{B, C\}$ since it is not in L but all its subsets are. The whole negative border is

$$Bd^-(L) = \{\{B,C\}, \{B,F\}, \{D\}, \{E\}\}.$$

The intuition behind the concept is that, given a collection L of itemsets that are frequent, the negative border contains the "closet" itemsets that could be frequent, too.

Definition 3. *The closed set $CS(L)$ of a collection of itemsets L is defined as follows. $CS(L) \equiv L \cup Bd^-(L)$.*

4.2 Secure Building Blocks

In this subsection, we review some secure protocols that will be used later in our algorithms.

Protocol 1. Homomorphic Secure Sum Protocol
1: Party 1 generates public key pair with a homomorphic encryption. It then shares the public key to all parties.
2: Party 1 calculates: $s_1 = E(v_1)$ where E(.) is the encryption operation. Then site 1 sends s_1 to site 2.
3: Each party i, where $2 \leq i \leq m$, gets s_{i-1} from party $i - 1$ and calculates: $s_i = s_{i-1} \cdot E(v_i)$
4: Party m send s_m to party 1
5: Party 1 decrypts s_m using his private key and shares the result to all parties.

4.3 Secure Sum

Secure Sum allows parties to securely compute the sum of data values from many parties. Assume that party i holds a value v_i. They want to calculate $v = \sum_{l=1}^{m} v_l$, where $0 \leq v \leq n$. In SMC method, there are two basic protocols that are allow to calculate a sum of many values from different sites in a secure manner.

Homomorphic encryption can be used to compute secure sum as shown in Protocol 1. This protocol is secure unless there is a collusion between the first party, who is private key keeper, and any other party.

Another method to securely compute sum of many values was proposed by Clifton et al. [6].

4.4 Secure Comparison

Let take the famous problem about two millionaires, named Alice and Bob. They want to know who is richer without revealing their money. The problem is originally presented by A. Yao [20]. He used secure comparison protocols to solve the problem. Yao then improved his method [21] to obtain a protocol that takes a linear complexity. Ioannidis then introduced a protocol that could perform secure comparison in logarithm time [10]. For the details of protocol, please refer to [10, 20, 21].

4.5 The Incremental Large Itemset Mining

In this section, we review a fast algorithm for incremental update of association rule mining proposed by Thomas et al. in [16]. The algorithm is done with two support algorithms: Apriori Generator and Negative Border Generator. Then the main algorithm Update-Large-Itemset is reviewed.

Apriori Generator. Protocol 2 demonstrates the appriori generation to produce k-large candidate itemsets when it has the set of $(k - 1)$-large frequent itemsets. It takes an argument L_{k-1}, set of $(k - 1)$-large itemsets and returns candidate k-large itemset. The correctness of protocol is proved in [1]. As we can see from the protocol, it can be done by any party without compromising

---

Protocol 2. Apriori-Gen

Require: L_{k-1}
Ensure: C_k
 1: **for each** p and $q \in L_{k-1}$ **do**
 2:   **if** $p.item_1 = q.item_1, ..., p.item_{k-2} = q.item_{k-2}$ and $p.item_{k-1} < q.item_{k-1}$
      then
 3:     Insert itemset $p.item_1, ..., p.item_{k-1}, q.item_{k-1}$ into C_k
 4:   **end if**
 5: **end for**
 6: **for each** $c \in C_k$ **do**
 7:   **if** some $(k-1)$-subset of $c \notin L_{k-1}$ **then**
 8:     Delete c from C_k
 9:   **end if**
10: **end for**

Protocol 3. Neg-Border-Gen

Require: L
Ensure: $L \cup Bd^- L$
 1: Split L into $L_1, L_2, ..., L_n$ where n is the size of the largest itemset in L
 2: **for each** $k = 1, 2, ..., n$ do **do**
 3:   Compute C_{k+1} using $apriori - gen(L_K)$
 4: **end for**
 5: $L \cup Bd^- L = \bigcup_{i=2,3,...,n+1} C_k \cup I_1$ where I_1 is the set of 1-itemsets

---

data privacy. The reason is that by default, all parties hold $L = \bigcup_{k=1..n} L_k$ at the end of the protocol. Hence, there is no need to apply privacy preserving techniques for this protocol.

Negative Border Generator. Protocol 3 shows the negative border generation. The input is L, a collection of all large itemset. The output is the negative border of L along with L itself. The correctness of this protocol can be found in [16]. Since the protocol can be performed by any party, there is no need to consider privacy issues. Hence, it can directly be applied into privacy preserving data mining versions. Figure 1 show the relationship between L^{DB}, L^{db} and $Bd^-(L^{DB})$.

Update Large Itemsets. Thomas *et al.* proposed an efficient algorithm to incremental update large itemset in [16]. The algorithm is presented in Protocol 4. The correctness proof of this algorithm can be found in [16]. The algorithm includes the following parts:

– **Part 1:** Compute large itemsets for new data at all sites (Step 1). All new parties or parties with data changed can used FUP algorithm to compute large itemset on new data.

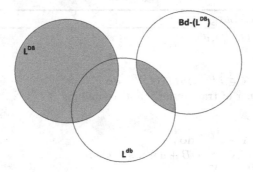

Fig. 1. The relationship of L^{DB}, L^{db} and $Bd^-(L^{DB})$.

- **Part 2:** Update large itemsets for all sites (Step 2–15). These steps are used to update new large itemsets or delete ones that are no longer large in new data.
- **Part 3:** Compute negative border of large itemsets and if there are some new itemsets, scan old datasets to update final large itemsets (Step 16–28).

Security Analysis. In this part, we analyze the privacy-preserving issues of the Prtocol 4. Steps 2–4 clearly breach privacy information: new parties know all itemsets in $L^{DB} \cup Bd^-(L^{DB})$ and old parties know the support count in new parties for each itemset $X \in L^{DB} \cup Bd^-(L^{DB})$.

5 The Proposed Incremental Privacy-Preserving Association Rule Mining

In this section, we use the tools described above to construct an incremental protocol that preserves the privacy of every site. Our algorithm follows the general method of the Update-Large-Itemset algorithm as reviewed in Sect. 4.5.

The protocol is presented in Protocol 5. It includes the following parts:

- Compute large itemsets in new sites only
- Compute Negative Border for old and new sites
- Update large itemsets for all sites.

In the next sections, we sequentially discuss in detail for each step.

5.1 Computing Large Itemsets in New Sites

In Step 1, all new sites simply apply the algorithm in [11]. The first site keeps a list of random numbers x_p, where $p = 1, ..., |L^{db}|$. The last site holds: $S = \sum_{i=1}^{r}(X.sup_i - |DB_i| \times s\%) + x_p$. To check if itemset X is large, the first and the last site only need to compare S with x_p using Yao's protocol [21]. After this step, every new site holds the large itemset L^{db}.

Protocol 4. Update-Large-Itemset

Require: $L^{DB}, Bd^-(L^{DB}), db$.
Ensure: L^{DB+}

1: Compute L^{db}
2: **for** each itemset $X \in L^{DB} \cup Bd^-(L^{DB})$ **do**
3:    $t_{db}(X)$ = number of transactions in db containing X
4: **end for**
5: $L^{DB+} = \emptyset$
6: **for** each itemset $X \in L^{DB}$ **do**
7:    **if** $t_{DB}(X) + t_{db}(X) \geq (DB + db) \times s\%$ **then**
8:       $L^{DB+} = L^{DB+} \cup X$
9:    **end if**
10: **end for**
11: **for** each itemset $X \in L^{db}$ **do**
12:    **if** $X \notin L^{DB}$ and $X \in Bd^-(L^{DB})$ and $t_{DB}(X) + t_{db}(X) \geq (DB + db) \times s\%$ **then**
13:       $L^{DB+} = L^{DB+} \cup X$
14:    **end if**
15: **end for**
16: **if** $L^{DB} \neq L^{DB+}$ **then**
17:    $Bd^-(L^{DB+})$=negative-border-gen(L^{DB+}))
18: **else**
19:    $Bd^-(L^{DB+}) = Bd^-(L^{DB})$
20: **end if**
21: **if** $L^{DB} \cup Bd^-(L^{DB}) \neq L^{DB+} \cup Bd^-(L^{DB+})$ **then**
22:    $S = L^{DB+}$
23:    **repeat**
24:       Compute $S = S \cup Bd^-(S)$
25:    **until** S does not grow
26: **end if**
27: $L^{DB+} = \{X \in S | sup(X) \geq s\}$
28: $Bd^-(L^{DB+})$ = negative-border-gen(L^{DB+}))

5.2   Compute Negative Border

After Step 1, all new sites hold L^{db} while all old sites hold $Bd^-(L^{db}) \cup L^{db}$. Applying Secure Union Set proposed by Tassa in [15], all sites can securely compute union set $L = L^{db} \cup Bd^-(L^{db}) \cup L^{db}$.

5.3   Updating Large Itemsets for All Sites

Steps 3–10, all sites compute large itemsets. Details are as follows.

- Step 4: New sites compute support count of itemsets using Secure Sum building blocks. There is no privacy leakage here as only new sites working together.
- Step 5: Old sites compute support count of itemsets using Secure Sum building blocks. Again, this step is privacy preserving as it relates to old sites only.
- Step 6: Using Secure Sum and Secure Comparison building blocks, one of the old sites and one of the new sites can easily check if an itemset is large.

Protocol 5. Incremental Privacy Preserving Large Itemset Mining

---

Require: $L^{DB}, Bd^-(L^{DB})$ from n old sites. $DB_1, DB_2, ..DB_r$ from r new sites.
Ensure: L^{DB+}: The frequent itemset of $(n + r)$ sites.
 1: All r new sites compute L^{db} using protocol in [11]
 2: Compute $L = L^{db} \cup Bd^-(L^{db}) \cup L^{db}$ using Secure Union Set
 3: **for** each itemset $s \in L$ **do**
 4:    All new sites compute $t_{db}(s)$, the support count of the itemset s in new sites
       using Secure Sum
 5:    All old site compute $t_{DB}(s)$, the support count of the itemset s in old sites using
       Secure Sum
 6:    One of the new sites and one of the old sites together use Secure Sum and Secure
       Comparison to check if s is a large itemset.
 7:    **if** s is large **then**
 8:        $L^{DB+} = L^{DB+} \cup s$
 9:    **end if**
10: **end for**
11: One of the sites compute Negative Border: $L^{DB+} \cup Bd^-(L^{DB+})$
12: **if** $L^{DB} \cup Bd^-(L^{DB}) \neq L^{DB+} \cup Bd^-(L^{DB+})$ **then**
13:    $L^{DB+} = \{X \in S | sup(X) \geq s\}$
14: **end if**

- Steps 7–10: The itemset is put into L^{DB+}, the set of all large itemset in both
  new and old data.
- Step 11: Compute Negative Border for L^{DB+} using Protocol 3.
- Step 12: One of the old site checks if the Negative Border of L^{DB} and L^{DB+}
  are the same. If they are the same, it means that all candidate new large
  itemset are in the Negative Border of L^{DB}, where their support counts have
  been computed before. Hence, there is no need to scan old datasets again. On
  the other hand, if the Negative Border of L^{DB} and L^{DB+} are different, there
  is a need to scan old datasets on the old sites once more time to compute
  support counts for new candidate itemsets. The correctness of this statement
  is similar to that of Protocol 4, which is proved by Thomas *et al.* in [16].

Theorem 1. *Protocol 5 is correct, i.e., it correctly returns all itemsets that are
frequent in combined data from old and new sites.*

Proof. Since Protocol 5 is derived based on Protocol 4, which is a non-secure
version, the protocol is correct if Protocol 4 is correct. We have known that
Protocol 4 is correct as it is proved in Theorem 1 in [16].

Thus we can safely conclude that Protocol 5 is correct, i.e., it generates all
large itemsets. □

Theorem 2. *Protocol 5 is secure in semi-honest model.*

Proof. To prove that the protocol is secure, we need to prove that each step is
secure.

- Step 1: This step is secure as it uses a secure protocol proposed by Kantarcioglu [11].
- Step 2: This step can be done by one of the new site and one of the old site using an efficient secure set union proposed by Tassa in [15]. Thus we can say that this step is secure.
- Step 4: Computing sum between new site is secure by using Secure Sum.
- Step 5: Similar to Step 4, this step is also secure with Secure Sum building blocks.
- Step 6: Since this step is to check if an itemset is frequent using Secure Sum and Secure Comparison building blocks, this step is secure too.
- Step 8: This step can securely done using Secure Set Union as in Step 2.
- Steps 11–12: Same as Step 8.
- Step 13: When it needs to scan old datasets again, we can apply the protocol by Kantarcioglu [11] to securely compute support count of itemsets and check if they are frequent. This step is thus secure too.

All steps in the protocol are secure. Then we can conclude that Protocol 5 is secure.   □

6   Experiments

Section 6.1 describes parameters to generate synthesis datasets for experiments. Section 6.2 demonstrates the two experiments set to be conducted. And in Sect. 6.3, we discuss about the experiments' results.

6.1   Generating Synthetic Data

We have generated synthetic data using the same techniques as in [1,15]. Table 1 presents a list of parameters to generate synthesis data. Except parameter m for number of sites, other parameters were used in previous work such as [1,3,11,14–16].

Table 1. Parameters to generate synthetic data.

Parameter	Description	Value
m	Number of sites	10
N	Number of transactions per site	500,000
L	Number of items	1,000
T	Mean size of a transaction	10
I	Mean size of maximal potentially large itemsets	4

6.2   Experimental Setup

We assume that k sites are old sites, i.e., they have completed running association rule mining using the protocol presented in [15]. Now the other $10 - k$ sites are new and want to join the mining task to get final results. We will compare the running time of two approaches: (i) All 10 sites have to run again protocol in [15], called **TASSA14** method. (ii) All 10 sites apply our incremental method as in Protocol 5, called **INCRE** method.

We have conducted two experiment sets as follows.

– We fix $k = 5$, i.e., 5 old sites and 5 new sites. The support threshold varies from 0.5 % to 2 %. The running time of two approaches have been computed.
– We fix the support threshold at 2 %. The number of old sites varies from 2 to 8. Hence, the number of new sites changes from 8 to 2. We also measure the running time to compare. Note that running time of **TASSA14** remains unchanged although the number of new sites is changed. The reason is that the protocol always runs on data of all sites.

We have implemented the protocol in [15] and our proposed protocol in Java. Each site is running on a Windows 7 64-bit OS with features: 8 GB of RAM, Intel Xeon E5-1620 3.6 GHZ of CPU. All sites are connected to each other via Gigabit ethernet cable.

6.3   Experimental Results

Figure 2 presents the total running time of two approaches: **TASSA14** method and our **INCRE** method when support threshold changes. From the results, we can see that the running time of two methods are decreasing when we increase support threshold. This is a known result as when support thresholds are bigger, the number of candidate itemsets are smaller. Thus both protocols will have to run less iterations. The interesting result in this experiment is that **INCRE** takes less time than **TASSA14** to complete the large itemset mining. This can be explained as follows. While **TASSA14** has to run on datasets of all 10 sites, **INCRE** runs on datasets of 5 new sites. Our protocol then makes use of results from old 5 sites (which is generated before) along with results from 5 new sites to compute the final large itemsets. Experimental results show that our method can reduce total running time about 50 % comparing with thaose of **TASSA14**.

Figure 3 shows the total running time of the two methods when the number of old site changes (total number of sites is still 10). If there are 2 old sites, then there are 8 new sites and so on. We can see that total running time of **TASSA14** remains unchanged. The total running time of our protocol is decreasing when the number of old sites increase (or number of new sites decrease). This can be explained as follows. When there are many old sites, our protocols makes use of old results from those sites and hence cut down the time to run on them again. The more old sites there are, the more time our protocol can save. If there are less new sites, then the protocols takes less time to access new datasets and compute new large itemsets. In the real world, there is normally less new

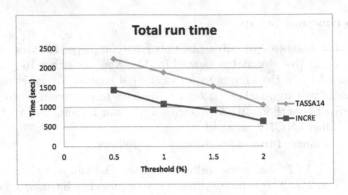

Fig. 2. Total run time of **TASSA14** versus **INCRE** when changing support threshold. Fixed 5 old sites and 5 new sites.

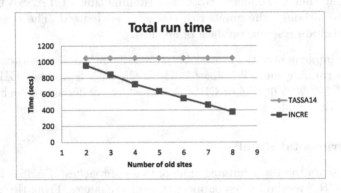

Fig. 3. Total run time of **TASSA14** versus **INCRE** when changing the number of old sites. Threshold fixed at 2 %.

sites than old sites. Our protocol thus expects to much outperform **TASSA14**. In this experiment, when there is 8 old sites and 2 new sites, our protocol can reduce total running time at 70 %.

7   Summary

In this paper, we have proposed a novel incremental protocol for secure mining of association rules in horizontally distribution. The protocol improves the fast incremental algorithm in [16] in term of privacy. One of the main improvements is making use of Negative Border in secure way to boost the performance of protocol. Thanks to this feature, our protocol outperforms the latest privacy preserving association rule mining proposed by Tassa in [15].

As we know that in real world, datasets can be deleted or some parties want to leave the mining process, a protocol to adapt with this changes is expected. In the future work, we will apply the concept of our protocol to deal with "deletion" of data.

Acknowledgments. This research was supported by Project SMI-2014-MA-05 awarded by Maritime Institute, Nanyang Technological University.

References

1. Agrawal, R., Srikant, R.: Fast algorithms for mining association rules. In: Bocca, J.B., Jarke, M., Zaniolo, C. (eds.) Proceedings of 20th International Conference on Very Large Databases, VLDB, pp. 487–499. Morgan Kaufmann (1994)
2. Ayan, N., Tansel, A., Arkun, E.: An efficient algorithm to update large itemsets with early pruning. In: Proceedings of the Fifth ACM SIGKDD International Conference on Knowledge Discovery and Data Mining, pp. 287–291. ACM (1999)
3. Cheung, D., Han, J., Ng, V., Fu, A., Fu, Y.: A fast distributed algorithm for mining association rules. In: Fourth International Conference on Parallel and Distributed Information Systems, pp. 31–42. IEEE (1996)
4. Cheung, D., Han, J., Ng, V., Wong, C.: Maintenance of discovered association rules in large databases: an incremental updating technique. In: Proceedings of the Twelfth International Conference on Data Engineering, pp. 106–114. IEEE (1996)
5. Cheung, D.W., Lee, S.D., Kao, B.: A general incremental technique for maintaining discovered association rules. In: Proceedings of the Fifth International Conference On Database Systems For Advanced Applications, pp. 185–194 (1997)
6. Clifton, C., Kantarcioglu, M., Vaidya, J., Lin, X., Zhu, M.Y.: Tools for privacy preserving distributed data mining. SIGKDD Explor. Newsl. **4**(2), 28–34 (2002)
7. Duan, Y., Canny, J.F., Zhan, J.Z.: Efficient privacy-preserving association rule mining: p4p style. In: CIDM, pp. 654–660. IEEE (2007)
8. Goldreich, O.: Secure multi-party computation. Manuscript (2002)
9. Han, S., Ng, W.-K.: Privacy-preserving genetic algorithms for rule discovery. In: Song, I.-Y., Eder, J., Nguyen, T.M. (eds.) DaWaK 2007. LNCS, vol. 4654, pp. 407–417. Springer, Heidelberg (2007)
10. Ioannidis, I., Grama, A.: An efficient protocol for yao's millionaires' problem. In: Proceedings of the 36th Annual Hawaii International Conference on System Sciences, p. 6, January 2003
11. Kantarcioglu, M., Clifton, C.: Privacy-preserving distributed mining of association rules on horizontally partitioned data. IEEE Trans. Knowl. Data Eng. **16**(9), 1026–1037 (2004)
12. Lee, C., Lin, C., Chen, M.: Sliding-window filtering: an efficient algorithm for incremental mining. In: Proceedings of the tenth international conference on Information and knowledge management, pp. 263–270. ACM (2001)
13. Mannila, H., Toivonen, H.: On an algorithm for finding all interesting sentences. In: Cybernetics and Systems, Volume II, The Thirteenth European Meeting on Cybernetics and Systems Research. Citeseer (1996)
14. Park, J.S., Chen, M.-S., Yu, P.S.: An effective hash-based algorithm for mining association rules. In: Proceedings of the ACM SIGMOD International Conference on Management of Data, SIGMOD 1995, pp. 175–186. ACM, New York (1995)
15. Tassa, T.: Secure mining of association rules in horizontally distributed databases. IEEE Trans. Knowl. Data Eng. **26**(4), 970–983 (2014)
16. Thomas, S., Bodagala, S., Alsabti, K., Ranka, S.: An efficient algorithm for the incremental updation of association rules in large databases. In: Knowledge Discovery and Data Mining, pp. 263–266 (1997)
17. Vaidya, J., Clifton, C.: Secure set intersection cardinality with application to association rule mining. J. Comput. Secur. **13**(4), 593–622 (2005)

18. Veloso, A., Meira Jr., W., De Carvalho, M., Pôssas, B., Parthasarathy, S., Zaki, M.: Mining frequent itemsets in evolving databases. In: SIAM International Conference on Data Mining (2002)
19. Wong, W.K., Cheung, D.W., Hung, E., Liu, H.: Protecting privacy in incremental maintenance for distributed association rule mining. In: Washio, T., Suzuki, E., Ting, K.M., Inokuchi, A. (eds.) PAKDD 2008. LNCS (LNAI), vol. 5012, pp. 381–392. Springer, Heidelberg (2008)
20. Yao, A.C.: Protocols for secure computations. In: SFCS 1982: Proceedings of the 23rd Annual Symposium on Foundations of Computer Science, pp. 160–164. IEEE Computer Society, Washington (1982)
21. Yao, A.C.: How to generate and exchange secrets. In: Proceedings of the Annual IEEE Symposium on Foundations of Computer Science, pp. 162–167 (1986)

Differentially Private Multi-task Learning

Sunil Kumar Gupta$^{(\boxtimes)}$, Santu Rana, and Svetha Venkatesh

Center for Pattern Recognition and Data Analytics, Deakin University,
Geelong 3216, Australia
{sunil.gupta,santu.rana,svetha.venkatesh}@deakin.edu.au

Abstract. Privacy restrictions of sensitive data repositories imply that
the data analysis is performed in isolation at each data source. A prime
example is the isolated nature of building prognosis models from hospital
data and the associated challenge of dealing with small number of samples in risk classes (e.g. suicide) while doing so. Pooling knowledge from
other hospitals, through multi-task learning, can alleviate this problem.
However, if knowledge is to be shared unrestricted, privacy is breached.
Addressing this, we propose a novel multi-task learning method that preserves privacy of data under the strong guarantees of differential privacy.
Further, we develop a novel attribute-wise noise addition scheme that
significantly lifts the utility of the proposed method. We demonstrate
the effectiveness of our method with a synthetic and two real datasets.

1 Introduction

Privacy matters. Tough legislations are in place to safeguard the privacy of
individual data repositories, resulting in a large set of disconnected data islands
with no means of connection between them. However, as data mining researchers
we believe that knowledge is everywhere and the true potential of data will be
unlocked when these disparate data islands are appropriately bridged - a pursuit
remains unrealized in the presence of privacy restrictions.

Consider healthcare as an example. Modern healthcare facilities are equipped
with Electronic Medical Records systems that capture detailed information
about patients as they access hospital facilities. The value of such information is
immense in creation of accurate prognosis models, central to efficient and appropriate delivery of care. However, we often encounter diseases that have *small
number of samples in risk classes*, e.g. suicide is rare in populations. The prognosis model built in such situations may result in poor performance. Pooling
knowledge across hospitals via *multi-task learning* framework can alleviate this
problem, but privacy-protecting regulatory frameworks control access to sensitive data across hospital jurisdictions. Similar situations arise in other areas,
e.g. building spam filters in a collaborative yet privacy-preserving manner etc.
*Therefore, there is opportunity to develop privacy preserving multi-task learning
models that provides strong guarantees on privacy protection.*

Early work on privacy preserving data analysis has used a variety of methods, e.g. query restriction [1], anonymization of information [2], secure multi-
party function evaluation [3], data/output perturbations [4] etc. Of these, query

© Springer International Publishing Switzerland 2016
M. Chau et al. (Eds.): PAISI 2016, LNCS 9650, pp. 101–113, 2016.
DOI: 10.1007/978-3-319-31863-9_8

restriction provides limited utility [5], anonymization may reveal sensitive data in presence of auxiliary information [6] and secure multi-party function evaluation may not provide statistical guarantees. Recently, differential privacy has emerged as a framework for privacy preserving information disclosure with strong theoretical guarantees [7]. It ensures that the answer to a statistical query is not significantly different between two datasets that differ at most in one instance. A major strength of differential privacy is its ability to provide graded levels of privacy by specifying a leakage parameter ε, giving rise to the name ε-differential privacy. Differential privacy has been applied to many data mining areas [7,8]. The closest work to our problem of building prediction models is the output perturbation approach due to [7], who suggested that differentially private algorithms can be constructed by calibrating the standard deviation of a Laplacian noise according to the "sensitivity" of a function involved in the algorithm. This idea was followed by Chaudhuri et al. to build differentially private empirical risk minimization models [9]. However, applicability of these models is limited to only *single-task learning* scenarios.

Current methods leverage collective knowledge across prediction problems (or tasks) via multi-task learning [10,11], building prediction models where inter-task knowledge transfer is achieved via some form of joint modeling. Existing multi-task learning models, however, are not equipped to satisfy privacy requirements as they require unrestricted access to sensitive data [11] or derived statistics [10]. A recent state-of-the-art multi-task learning method is MTRL [12], which provides a flexible way of sharing knowledge across tasks by using a covariance matrix to model task relationships. As a result, it is able to exploit knowledge from tasks that have varying degree of relatedness - a crucial property when dealing with real world data.

Limited work exist on privacy preserving multi-task learning. Mathew and Obradovic [13] construct a distributed Id3-based decision tree for predicting hospitalization risk from multi-hospital data. Although no data is exchanged between the hospitals, leakage on privacy may occur due to exchanging unperturbed statistics. Pathak et al. [14] propose a differentially private multi-task learning via averaging classifiers from multiple sources using secure multiparty communication. This method has two drawbacks. (1) Averaging classifiers assumes that tasks are strongly correlated, and (2) the level of noise is calibrated with respect to the *smallest* source, resulting in *high* model perturbation. This seriously brings down the utility of the algorithm. *Therefore, the opportunity to develop a differentially private multi-task learning model* is open.

Taking this opportunity, we propose a novel multi-task learning that preserves privacy of individuals in participating sources, under the strong guarantee of differential privacy. The proposed model infuses privacy into the MTRL model [12]. This delivers strong privacy preserving property to a state-of-art multi-task learning model facilitating seamless sharing of knowledge without sharing data between the participating sources. In case of healthcare, this means that hospitals across the world can improve their prognosis models by leveraging their mutual knowledge and thus derive best practice to revolutionize healthcare.

Following sensitivity based approach of Dwork et al. [7], we derive the sensitivity for the proposed multi-task learning model and use it to calibrate the level of noise used for differential privacy. We provably show that the proposed scheme satisfies ε-differential privacy where it is possible for a source (e.g. hospital) to control its privacy requirements via its own ε parameter. Adding calibrated noise to task parameters helps us in securing privacy, however, it comes at the cost of reduced performance (or utility). Addressing this problem, we propose a novel scheme of attribute-wise noise addition that exploits the information content of attributes and *reduces* the overall noise. We demonstrate that this scheme can significantly lift the performance of the proposed technique. Using a synthetic dataset, we illustrate the behavior of the proposed models and validate their effectiveness using *two* real world problems: predicting cancer mortality and designing personalized spam filter.

2   Preliminaries

Differential Privacy. Differential privacy is a privacy preserving framework proposed by Dwork et al. [7]. This framework defines a notion of privacy for a learning algorithm $\mathscr{A}$. The algorithm $\mathscr{A}$ satisfies differential privacy if likelihood of its output for two datasets that differ at most by one instance are close. Due to this closeness, an adversary can not infer anything significant about the differing instance by using the algorithm output. The closeness of the likelihoods is characterized by a "leakage" parameter ε, giving rise to the name ε-differential privacy.

Definition 1 [7]: *An algorithm $\mathscr{A}$ is said to satisfy ε-differential privacy if for any two datasets $\mathscr{D}$ and $\mathscr{D}'$ that differ by at most one instance, and all $S \subseteq Range\,(\mathscr{A})$,*

$$\exp\left(-\varepsilon\right) \leq \frac{P\left(\mathscr{A}\left(\mathscr{D}\right) \in S\right)}{P\left(\mathscr{A}\left(\mathscr{D}'\right) \in S\right)} \leq \exp\left(\varepsilon\right) \tag{1}$$

where $\mathscr{A}\left(\mathscr{D}\right)$ and $\mathscr{A}\left(\mathscr{D}'\right)$ are the outputs of $\mathscr{A}$ on datasets $\mathscr{D}$ and $\mathscr{D}'$ respectively.

Sensitivity. The sensitivity of a function f is the maximum change in its output due to any single data instance. A formal definition of sensitivity is provided below:

Definition 2 [7]: *The sensitivity of a function $f : D \rightarrow \mathbb{R}^M$ is defined as*

$$S\left(f\right) = \max_{\mathscr{D},\mathscr{D}'} \|f\left(\mathscr{D}\right) - f\left(\mathscr{D}'\right)\|$$

for all datasets $\mathscr{D}$ and $\mathscr{D}'$ that differ by at most one instance. Dwork et al. [7] showed that ε-differential privacy is satisfied by an algorithm if i.i.d. Laplacian noise with standard deviation $S\left(f\right)/\varepsilon$ is added in each co-ordinate of the output vector before its release.

Strong Convexity. The strong convexity property is used to derive our proposed model. We provide the definition of strong convexity in the following.

Definition 3. *A twice continuously differential function $C(f)$ is called strongly convex with parameter $\mu > 0$ iff the following inequality holds for all f in its domain*

$$\nabla^2 C(f) \succ \mu I, \tag{2}$$

where $\succ$ means that $\nabla^2 C(f) - \mu I$ is positive semi-definite.

3   Privacy Preserving MTL

Let us assume we have T_0 tasks, indexed as $t = 1, \ldots, T_0$. For the t-th task, we denote the training set as $\mathscr{D}_t = \{(\mathbf{x}_{ti}, y_{ti})\}_{i=1}^{N_t}$ where $\mathbf{x}_{ti} \in \mathbb{R}^M$ is a M-dimensional feature vector and y_{ti} is the target, usually real-valued for regression and binary-valued for binary classification problems. Let β_t denote the weight vector for the task t, we also refer to this as *task parameter*. Collectively, we denote the data of t-th task by $\mathbf{X}_t = (\mathbf{x}_{t1}, \ldots, \mathbf{x}_{tN_t})^T$ and $\mathbf{y}_t = (y_{t1}, \ldots, y_{tN_t})^T$ and all the task parameters as $\beta = (\beta_1, \ldots, \beta_{T_0})$. When tasks differ in some of the features, a common feature list can be obtained via their union.

The multi-task learning literature is full of sophisticated models where the aim is to jointly model multiple tasks towards improved average prediction performance for all the tasks. In this paper, we use a multi-task learning model that learns relationship of tasks via a covariance matrix and uses it for joint modeling [12]. Although we have chosen this model to build a privacy preserving variant, one can use the technique described in this paper to many other multi-task learning models provided these models minimize a convex loss function. The results for non-convex models are more involved and out of scope of this paper.

The proposed multi-task learning model minimizes the following objective function

$$\min_{\beta, \Omega} \sum_t \frac{||\mathbf{X}_t \beta_t + b_t \mathbf{1} - \mathbf{y}_t||^2}{N_t} + \lambda_1 \mathrm{Tr}\left(\beta \beta^T\right) + \lambda_2 \mathrm{Tr}\left(\beta \Omega^{-1} \beta^T\right), \quad \text{s.t. } \Omega \succeq 0, \ \mathrm{tr}(\Omega) = 1 \tag{3}$$

where b_t is the bias parameter of the t-th task and the notation $\mathbf{1}$ denotes a vector of all ones. *We refer to the above cost function as $C(\beta, \Omega)$.* Although, in this paper, we use the square loss, it is possible to extend this formulation for logistic loss. Similarly, extensions to multi-class classification is straight-forward. We take the above model and build its privacy preserving variant, which protects the data from being reversed engineered by an adversary from model parameters.

Since the cost function of 3 is jointly convex in β and Ω along with the constraints, we can find unique solution. Our approach is to optimize β for a fixed Ω and then optimize Ω given β. This leads to an iterative solution.

For square loss, task parameter β_t given Ω can be learnt in a closed form. This is done by setting the derivative of $C(\beta, \Omega)$ with respect to β_t to zero, leading to the following linear equation in β_t

---

Algorithm 1. The proposed Private-MTL

1: **Input**: Multi-task data $\{\mathbf{X}_t, \mathbf{y}_t\}_{t=1}^{T_0}$, parameters λ_1, λ_2, ε.
2: **Output**: Task parameters $\beta_{1:T_0}$ and matrix Ω.
3: **Initialization:** For initialization, learn task parameters using single task learning (STL), let us assume the task parameter for task t using STL is β_t, computed locally.
4: compute sensitivity for task t as $S_t = \frac{2}{N_t \Lambda_t}$ where $\Lambda_t = \lambda_1 + \lambda_2 \Omega^{-1}(t,t)$ and N_t is the number of instances in t-th task.
5: sample η_t from the density function: $p(\eta_t) \propto \exp\left(-\frac{\varepsilon}{S_t}||\eta_t||\right)$.
6: Add noise to task parameters as $\beta_t = \beta_t + \eta_t$.
7: **repeat**
8:    update task relationship matrix as $\Omega = \frac{(\beta^T \beta)^{1/2}}{\mathrm{Tr}\left((\beta^T \beta)^{1/2}\right)}$.
9:    solve β_t given Ω and other noisy $\beta_{t'}$, $t' \neq t$ using (4).
10:    set $\beta_t = \beta_t + \eta_t$ where η_t is sampled similar to step-5.
11: **until** convergence

$$\left[\left(\mathbf{X}_t^T \mathbf{X}_t\right)/N_t + \left(\lambda_1 + \lambda_2 \Omega^{-1}(t,t)\right)\mathbf{I}\right]\beta_t = \left(\mathbf{X}_t^T\left(\mathbf{y}_t - b_t \mathbf{1}\right)\right)/N_t - \lambda_2 \sum_{t' \neq t} \Omega^{-1}(t',t)\,\beta_{t'}. \quad (4)$$

As seen from this equation, for a fixed task relatedness Ω, learning task parameter of t-th task, i.e. β_t requires data from only its own task. The knowledge from *other tasks* is brought through their task parameters, i.e. $\beta_{t'}$ where $t' \neq t$. To have a solution that preserves privacy according to the ε-differential privacy, we follow the sensitivity method suggested by [7]. We compute sensitivity (S_t) of our objective function and add a noise vector (η_t) calibrated using this sensitivity to the task parameters. This method can be shown to guarantee the privacy of data instances from all the tasks. Using this method, the noisy β_t is given as

$$\beta_t = \beta_t + \eta_t, \; p(\eta_t) \propto \exp\left(-\frac{\varepsilon}{S_t}||\eta_t||\right). \quad (5)$$

where we slightly abuse the notation of β_t using it to denote the task parameters both *before* and *after* adding noise.

For a fixed noisy β, the matrix Ω can be learnt by minimizing $\mathrm{Tr}\left(\beta\Omega^{-1}\beta^T\right)$ subject to constraints $\Omega \succeq 0, \mathrm{tr}\,(\Omega) = 1$. Zhang et al. [12] show that the closed form solution of this optimization problem is given as $\Omega = \frac{(\beta^T \beta)^{1/2}}{\mathrm{Tr}\left((\beta^T \beta)^{1/2}\right)}$. We note that there is no need to add noise in Ω, as it is estimated from β, which is already noisy and privacy preserving. For all future references, we term this model as **Private-MTL**.

In spite of adding noise to β_t, the convergence of the optimization function in (3) is still guaranteed under the framework of stochastic optimization. The noisy perturbations (with mean zero) to β_t can be thought as updating β_t using a noisy gradient of the cost function, which is popular in stochastic optimization literature and known to converge [15]. Algorithm 1 provides a step-by-step summary of the proposed model.

Privacy Guarantees. We establish the conditions under which Algorithm 1 provide ε-differential privacy with respect to β. Proving differential privacy w.r.t. Ω is not necessary as by reverse engineering Ω, one can only reach to β, which is noisy and privacy preserving.

Theorem 1. *The task parameters $\beta_{1:T_0}$ learnt using Algorithm 1 preserves ε-differential privacy.*

Proof: The proof of the Theorem follows a similar sketch as the proof of Theorem 6 in Chaudhuri et al. [9]. Due to involvement of multi-task regularization, the sensitivity of the model, however, is different. Lemma 3 derives the sensitivity of our proposed model ($S_t = \frac{2L}{N_t \Lambda_t}$, where $\Lambda_t = \lambda_1 + \lambda_2 \Omega^{-1}(t,t)$ and loss function is assumed to be L-Lipschitz). This result, in combination with the sensitivity method of Dwork et al. [7] and output perturbation method of [9], establishes the theorem. For the sake of completeness, we provide a sketch of the proof here.

Let $\mathscr{D}$ and $\mathscr{D}'$ be any two datasets that differ only in n_t-th instance of task t. Further, let $\beta_t^{\mathscr{D}}$ and $\beta_t^{\mathscr{D}'}$ be the task parameters learnt using these datasets without any noise additions. Let β_t be the task parameter after noise addition. Then, for any β_t and dataset $\mathscr{D}$, we have $p(\beta_t|\mathscr{D}) \propto e^{-\frac{N_t \Lambda_t \varepsilon}{2L}(\|\beta_t - \beta_t^{\mathscr{D}}\|)}$, which leads to

$$\frac{p(\beta_t|\mathscr{D})}{p(\beta_t|\mathscr{D}')} = e^{-\frac{N_t \Lambda_t \varepsilon}{2L}\left(\|\beta_t - \beta_t^{\mathscr{D}}\| - \|\beta_t - \beta_t^{\mathscr{D}'}\|\right)} \tag{6}$$

where $p(\beta_t|\mathscr{D})$ and $p(\beta_t|\mathscr{D}')$ are the density function of the task parameter β_t given datasets $\mathscr{D}$ and $\mathscr{D}'$. Using triangular inequality and Lemma 3, we have

$$\|\beta_t - \beta_t^{\mathscr{D}}\| - \|\beta_t - \beta_t^{\mathscr{D}'}\| \leq \|\beta_t^{\mathscr{D}'} - \beta_t^{\mathscr{D}}\| \leq \frac{2L}{N_t \Lambda_t}.$$

Plugging this result in (6), we have $\frac{p(\beta_t|\mathscr{D})}{p(\beta_t|\mathscr{D}')} \geq e^{-\varepsilon}$. Due to symmetry in choosing $\mathscr{D}$ and $\mathscr{D}'$, we also have $\frac{p(\beta_t|\mathscr{D})}{p(\beta_t|\mathscr{D}')} \leq e^{\varepsilon}$, guaranteeing ε-differential privacy.

Lemma 2. *The cost function of (3) for task t is Λ_t-strongly convex with $\Lambda_t = \lambda_1 + \lambda_2 \Omega^{-1}(t,t)$.*

Proof: We note that $C(\beta, \Omega)$ is doubly-differentiable. In this light, consider the strong convexity condition in (2). To prove the lemma, we need to show: $\nabla^2_{\beta_t} C(\beta, \Omega) \succ (\lambda_1 + \lambda_2 \Omega^{-1}(t,t))\mathbf{I}$. The first derivative of the cost function in (3) is given as

$$\nabla_{\beta_t} C = \frac{\left(\mathbf{X}_t^T \mathbf{X}_t \beta_t - \mathbf{X}_t^T(\mathbf{y}_t - b_t\mathbf{1})\right)}{N_t} + \lambda_1 \beta_t + \lambda_2 \beta \Omega^{-1}(:,t). \tag{7}$$

Taking second derivative, we get the following result

$$\nabla^2_{\beta_t} C = \frac{1}{N_t}\mathbf{X}_t^T \mathbf{X}_t + \lambda_1 \mathbf{I} + \lambda_2 \Omega^{-1}(t,t)\mathbf{I}. \tag{8}$$

Clearly the matrix $\nabla^2_{\beta_t} C - (\lambda_1 + \lambda_2 \Omega^{-1}(t,t))\mathbf{I}$ is positive semi-definite as, for any $\mathbf{v}$, $\mathbf{v}^T(\mathbf{X}_t^T \mathbf{X}_t)\mathbf{v} = (\mathbf{X}_t \mathbf{v})^T(\mathbf{X}_t \mathbf{v}) \geq 0$. Therefore, the cost function $C(\beta, \Omega)$ for each β_t (i.e. task t) is Λ_t-strongly convex with $\Lambda_t = \lambda_1 + \lambda_2 \Omega^{-1}(t,t)$.

Lemma 3. Assuming bounded input ($||\mathbf{x}_{ti}|| \leq 1$) and L-Lipschitz assumption on the loss function, the sensitivity of $C(\beta, \Omega)$ for task t is at most $\frac{2L}{N_t \Lambda_t}$, where $\Lambda_t = \lambda_1 + \lambda_2 \Omega^{-1}(t, t)$.

Proof: To derive the sensitivity of $C(\beta, \Omega)$, consider two datasets $\mathscr{D}$ and $\mathscr{D}'$ that differ only in n_t-th instance of task t (denoted as $d_{t,n_t} = (\mathbf{x}_{t,n_t}, y_{t,n_t})$). Further let $G(\beta, \Omega) = C(\beta, \Omega)|_{\mathscr{D}'}$, $g(\beta) = C(\beta, \Omega)|_{\mathscr{D}} - C(\beta, \Omega)|_{\mathscr{D}'}$, $\beta_t^{\mathscr{D}} = \mathrm{argmin}_{\beta_t} C(\beta, \Omega)|_{\mathscr{D}}$, and $\beta_t^{\mathscr{D}'} = \mathrm{argmin}_{\beta_t} C(\beta, \Omega)|_{\mathscr{D}'}$. The sensitivity of β_t is given by $\max_{d_{t,n_t}} ||\beta_t^{\mathscr{D}} - \beta_t^{\mathscr{D}'}||$, which following a similar derivation as Lemma 7 and Corollary 8 in [9] can be shown to be at most $\frac{2L}{N_t \Lambda_t}$, where $\Lambda_t = \lambda_1 + \lambda_2 \Omega^{-1}(t, t)$.

Attribute-Wise Noise Addition. In above method, both ε and the sensitivity S_t are set identically for all attributes - noise is added isotropically. We aim to *reduce* the level of noise by exploiting attribute-specific properties. The main idea is that an attribute needs to be kept strictly private only when it is rich in information. For an attribute that does not carry much information, there is not much reason to make it private. For example, most patients in a cancer hospital will have chemotherapy, thus enforcing stringent privacy on 'whether someone has undergone chemotherapy or not' is unnecessary. For such attributes, we can relax the privacy constraint by setting parameter ε to a higher value. We do this by setting ε as a function of the attribute entropy ($\mathbb{H}_{ti}$). Entropy is a surrogate to capture the uniqueness of an attribute. In particular, we set the privacy level for the i-th attribute in task t as $\varepsilon_{ti} = \varepsilon_0 (1 + \kappa_f \exp(-\mathbb{H}_{ti}))$, where $\mathbb{H}_{ti}$ represents uncertainty of i-th attribute in task t and κ_f is an *"attribute-wise privacy scale parameter"*. The parameter κ_f decides the rate at which the privacy requirements are relaxed with decreasing attribute uncertainty. Depending on the level of entropy, ε_{ti} varies between $(\varepsilon_0, \varepsilon_0(1 + \kappa_f)]$. For continuous valued feature differential privacy can be used. Using attribute-wise privacy parameter ε_{ti}, for task t, the perturbation in i-th element of task parameter is given as

$$\beta_{ti} = \beta_{ti} + \eta_{ti}, \; p(\eta_{ti}) \propto \exp\left(-\frac{\varepsilon_{ti}}{S_t}|\eta_{ti}|\right). \tag{9}$$

The proof on privacy guarantee is similar to the proof in Theorem 3.1. Only difference is that we are now using independent Laplacian noise with different parameters for each attribute instead of using *i.i.d.* noise. In Algorithm 1, the step-5, step-6 and step-10 are appropriately replaced by Eq. (9).

4 Experiments

4.1 Experimental Setup

We experiment with a synthetic dataset and two real datasets.

We compare our Private-MTL with the following baselines: *(a) NonPrivate-STL:* In this algorithm, prediction weight vectors are learnt separately at each entity and released without privacy protection, *(b) NonPrivate-MTL:* In this algorithm, prediction weight vectors are learnt using multi-task learning but without privacy restriction, *(c) Private-STL:* In this algorithm, weight vectors of all entities are learnt separately and noise is added to preserve privacy, and *(d) MDP-AC:* In this algorithm [14], weight vectors are shared via secure multiparty computation and averaged to obtain a global classifier. The noise is added corresponding to the smallest dataset. For all the above baselines and the proposed models, the regularization parameters are learnt using cross-validation.

4.2   Experiments with Synthetic Dataset

We synthesize a multi-task learning dataset where tasks have various form of relationships: positive, negative and no relationship. Our aim is to show that our proposed model is able to estimate the task parameters (β_t) accurately even under privacy preserving restrictions. We create a total of 12 tasks with their task parameters defined in a 9-dim. space. We simulate three task groups putting the first 4 tasks in group-1, the next 4 tasks in group-2 and the last 4 tasks in group-3. We use different relatedness across task groups along different features. In particular, for the first 3 features, tasks in group-2 and group-3 are *positively* related, but *unrelated* to tasks in group-1. Similarly, for the next 3 features, tasks in group-1 and group-3 are *positively* related, but *unrelated* to tasks in group-2. Finally, for the last 3 features, tasks in group-1 and group-2 are

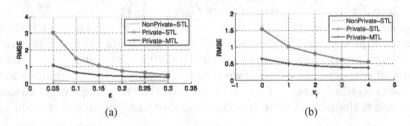

(a)   (b)

Fig. 1. Experimental results on Synthetic dataset: (a) Root Mean Square Error (RMSE) as a function of privacy parameter (ε), and (b) RMSE as a function of attribute-wise privacy scale parameter (κ_f) at $\varepsilon_0 = 0.1$.

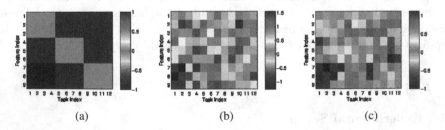

(a)   (b)   (c)

Fig. 2. Task parameters for Synthetic data experiments: (a) True, (b) Private-STL, and (c) Private-MTL. The task parameters shown are average of 50 run with $\varepsilon = 0.1$ and $\kappa_f = 1$; Between the task parameters obtained by Private-STL and Private-MTL, the latter resembles more to the true task parameters used for the synthesis of data.

negatively related, but unrelated to tasks in group-3. Overall relationship across tasks aggregated over all features becomes partial.

The i-th instance for t-th task, i.e. $\mathbf{x}_{ti}$ is generated from a 9-dim. multivariate normal distribution as $\mathbf{x}_{ti} \sim \mathcal{N}(\mathbf{0}, \mathbf{I})$. The outcome y_{ti} is generated as $y_{ti} = \beta_t^T \mathbf{x}_{ti} + e_{ti}$, $e_{ti} \sim \mathcal{N}(0, 0.1)$, where e_{ti} is a random noise. We generate 100 instances per task.

We randomly divide the data from each task into a 70 % training set and a 30 % test set. Figure 1a shows the average predictive performance as a function of the privacy parameter ε for the proposed Private-MTL and the two STL baselines. As seen from the figure, NonPrivate-STL provides the lowest RMSE error. This is because there is no noise addition in the task parameters. Out of the two privacy preserving algorithms, Private-MTL performs better than the Private-STL. This benefit comes from the ability of the Private-MTL successfully leveraging from the knowledge of the other tasks. At the stronger privacy requirements (i.e. for lower values of ε), the performance benefit is even more pronounced with Private-MTL performing two times better than Private-STL at $\varepsilon = 0.05$. Figure 1b shows similar plots as a function of κ_f when the global privacy parameter is set at $\varepsilon_0 = 0.1$. This plot clearly shows significant improvement in performance due to attribute-wise anisotropic noise addition. As the data is synthesized, we know the true task parameters and compare that with the recovered task parameters by different algorithms. Figure (2a–c) provide depiction of both the true and the recovered task parameters. It is evident that the task parameters recovered by Private-MTL resemble more closely to the true task parameters than that of by Private-STL.

4.3 Experiments with Real Dataset

Cancer Dataset. The Cancer dataset is obtained from a large regional hospital in Australia[1]. This cohort consists of 4,200 cancer patients who visited the hospital during 2010-2012. The data contains a variety of information such as patient demographics, diagnosis records in terms of ICD-10 codes, and procedure codes in terms of ACHI system. Features are extracted following [16], resulting in 683 features. The task involves 1 year mortality prediction. To simulate a *multi-hospital scenario*, we randomly divide the whole cohort into 5 separate cohorts that are assumed to be coming from hospitals of different sizes: a large hospital (LH) with 3000 patients, a medium hospital (MH) with 600 patients, and 3 small hospitals (SH) with 200 patients each. The *unequal* division reflects the typical real-world setting where several medium to small hospitals work together with a large hospital in the nearby city.

Multi-user Spam Dataset. The spam dataset is obtained from the *ECML-PKDD challenge* held in 2006. We use the test dataset from the Task B challenge. The dataset contains 15 users, each having 400 labeled emails. The emails are supplied as a term-document matrix with a dictionary size of around 150,000.

---

[1] Ethics approval obtained through University and the hospital – 12/83.

For each user spams constitute 50 % of the emails. The goal is to build a spam classifier for each user locally. Since the notion of spam emails is related across users, multi-task learning can be used to share classification knowledge across the users. However, emails being private in nature, we should aim to perform any knowledge sharing in a privacy preserving way.

Experimental Results

Cancer Dataset. Figure 3a shows the predictive performance averaged across all the hospitals as a function of privacy parameter ε on the Cancer dataset. Randomly selected 70 % data from each hospital is used for training and the rest for test. The experiment is performed 50 times and the average performance is reported along with respective standard error. As seen from the figure, the highest average Area Under the ROC Curve (AUC) over all the hospitals is achieved by the Private-MTL across the range of privacy parameter (ε) tested. Even at a stringent privacy requirement of $\varepsilon = 0.05$, Private-MTL achieves $\sim 5 \%$ higher AUC than the NonPrivate-STL. This indicates that the hospitals benefit by collaboration than building tools independently. This is significant since collaborating privately introduces noise in the estimation, yet improvement in prediction is achieved. Private-STL may not be a suitable benchmark as we may assume that learning independently at each hospital does not require the use of privacy preserving algorithms. However, it is still a useful comparator to illustrate the absolute gain achieved just because of multi-task learning. At $\varepsilon = 0.05$, the improvement achieved by the multi-task learning is more than 30 % over the Private-STL. As expected, the performance by both the Private-STL and Private-MTL improves with increasing ε (decreasing privacy requirement), however, the benefits obtained by using multi-task learning remains significant. Further, the performance of Private-MTL almost reaches the performance of Non-Private MTL. MDP-AC performs lower but somewhat closer to the Private-STL. As the data is originally from a single source averaging the weight vectors had the potential to work. However, the amount of noise is computed based on the smallest sized hospital. This has resulted in a higher noise, leading to a poor performance.

In the Cancer dataset, we have one large, one medium and three small sized hospitals. Table 1 shows the performance by different algorithms at different hospitals at $\varepsilon = 0.1$. We compare the performance between Private-MTL over the NonPrivate-STL. We see that the gain in performance by the Private-MTL is more for medium and small sized hospitals (average gain = 4.5 %) than the large hospital (gain = 1.5 %). This shows the higher need of multi-task learning for medium/smaller sized hospitals.

Figure 3b shows the use of attribute-wise privacy for the cancer dataset. Figure 4 shows the histogram of attribute-wise entropy across all the 5 hospitals. It clearly shows that there are many features which have low entropy. Figure 3b shows the predictive performance as a function of the attribute-wise privacy scale parameter κ_f when the global privacy parameter is set at $\varepsilon_0 = 0.1$. Average

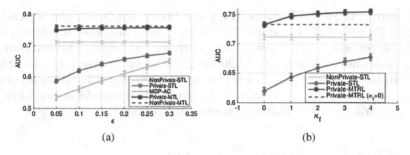

(a)   (b)

Fig. 3. (a) Average AUC of prediction on Cancer dataset as a function of the privacy parameter ε, and (b) average AUC of prediction on Cancer dataset as a function of the attribute-wise privacy parameter κ_f when $\varepsilon_0 = 0.1$. For both average performance is reported over 50 random training/test splits (std. errors shown as error-bars).

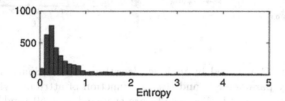

Fig. 4. Histogram of entropy across features over all the tasks (hospitals).

Table 1. Average AUC of prediction on Cancer dataset at different hospitals; a large hospital (LH) with 3000 patients, a medium sized hospital (MH) with 600 patients and three small hospitals (SH1-3) with 200 patients each at $\varepsilon = 0.1$. Performance is averaged over 50 random training/test splits. Standard errors are reported in parenthesis.

	AUC (std err)				
	LH	MH	SH1	SH2	SH3
NonPrivate-STL ($R1$)	0.800 (0.002)	0.683 (0.005)	0.750 (0.011)	0.713 (0.012)	0.612 (0.014)
Private-STL ($R2$)	0.791 (0.003)	0.632 (0.008)	0.576 (0.020)	0.573 (0.015)	0.527 (0.018)
Private-MTL ($R3$)	**0.815** (0.002)	**0.735** (0.005)	**0.787** (0.011)	**0.777** (0.010)	**0.657** (0.013)
$\Delta(R3 - R1)$	0.015	0.052	0.037	0.0.064	0.045

performance from 50 random splits of 70 % data for training and the rest for test is reported. The plot shows that AUC improves considerably with increasing κ_f. Private-MTL gains further 2.5 % in AUC at $\kappa_f = 4$.

Multi-user Spam Dataset. Figure 5a shows the average comparative predictive performance in terms of AUC on Spam dataset as a function of the privacy parameter ε. For each user, 70 % of the data is randomly selected for training and the rest for test. The average performance over 50 such splits are shown. As seen from the plot, the performance by all three privacy preserving algorithms are much worse than NonPrivate-STL at the stringent privacy requirement of $\varepsilon = 0.05$. However, they improved as the privacy requirement is lowered (ε

is increased). Private-MTL starts to become better at $\varepsilon \geq 0.2$, almost reaching up to NonPrivate-MTL. It implies that building spam filters collaboratively may perform better on average below a certain privacy restriction. This is a common phenomenon one will encounter while designing privacy preserving algorithms. Figure 5b shows the corresponding performance when $\varepsilon_0 = 0.1$ and the attribute-wise privacy parameter κ_f is varied. As expected, the performance improves when the attribute-wise anisotropic noise is introduced.

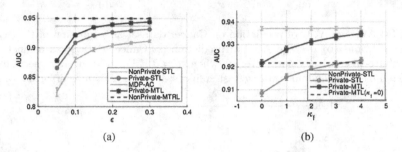

(a)   (b)

Fig. 5. Average AUC achieved by various algorithms on the Spam dataset: (a) as a function of privacy parameter ε, and (b) as a function of attribute-wise privacy scale parameter κ_f when $\varepsilon_0 = 0.1$. The results are averaged over 50 random training-test splits. Standard errors are shown as error bars.

5   Conclusion

We propose a novel multi-task learning model that preserves privacy of individuals at participating tasks under differential privacy. To lift the model's utility, we develop a novel attribute-wise noise addition scheme that adds anisotropic noise calibrated according to uncertainty of the attributes leading to reduced noise. Comparing with the state-of-art baselines on two real world datasets we demonstrate the effectiveness of our approach. In future, we will continue exploring connections between multi-task learning and privacy by extending differentially private random forest [17] to multi-task learning or extending model-agnostic multi-task learning [18] to a privacy-preserving variant.

References

1. Chin, F.Y., Ozsoyoglu, G.: Auditing and inference control in statistical databases. IEEE Trans. Softw. Eng. **8**(6), 574–582 (1982)
2. LeFevre, K., DeWitt, D.J., Ramakrishnan, R.: Incognito: Efficientfull-domain k-anonymity. In: SIGMOD, pp. 49–60. ACM (2005)
3. Ben-David, A., Nisan, N., Pinkas, B.: Fairplaymp: a system for securemulti-party computation. In: ACM CCS, pp. 257–266. ACM (2008)
4. Traub, J.F., Yemini, Y., Woźniakowski, H.: The statistical security of a statistical database. TODS **9**(4), 672–679 (1984)

5. Dinur, I., Nissim, K.: Revealing information while preserving privacy. In: PODS, pp. 202–210. ACM (2003)
6. Ganta, S., Kasiviswanathan, S., Smith, A.: Composition attacks and auxiliary information in data privacy. In: SIGKDD, pp. 265–273. ACM (2008)
7. Dwork, C., McSherry, F., Nissim, K., Smith, A.: Calibrating noise to sensitivity in private data analysis. In: Halevi, S., Rabin, T. (eds.) TCC 2006. LNCS, vol. 3876, pp. 265–284. Springer, Heidelberg (2006)
8. Vaidya, J., Clifton, C.W., Zhu, Y.M.: Privacy Preserving Data Mining, vol. 19. Springer Science & Business Media, New York (2006)
9. Chaudhuri, K., Monteleoni, C., Sarwate, A.D.: Differentially private empirical risk minimization. J. Mach. Learn. Res. **12**, 1069–1109 (2011)
10. Argyriou, A., Evgeniou, T., Pontil, M.: Convex multi-task feature learning. Mach. Learn. **73**(3), 243–272 (2008)
11. Saha, B., Gupta, S., Phung, D., Venkatesh, S.: Multiple task transfer learning with small sample sizes. In: Knowledge and Information Systems, pp. 1–28 (2015)
12. Zhang, Y., Yeung, D.-Y.: A convex formulation for learning task relationships in multi-task learning. In: Uncertainty in Artificial Intelligence, pp. 733–442 (2010)
13. Mathew, G., Obradovic, Z.: Distributed privacy preserving decision support system for predicting hospitalization risk in hospitals with insufficient data. In: ICMLA, vol. 2, pp. 178–183 (2012)
14. Pathak, M., Rane, S., Raj, B.: Multiparty differential privacy via aggregation of locally trained classifiers. In: NIPS, pp. 1876–1884 (2010)
15. Spall, J.C.: Introduction to Stochastic Search and Optimization: Estimation, Simulation, and Control, vol. 65. Wiley, Hoboken (2005)
16. Tran, T., Luo, W., Phung, D., Gupta, S., Rana, S., Kennedy, R.L., Larkins, A., Venkatesh, S.: A framework for feature extraction from hospital medical data with applications in risk prediction. BMC Bioinform. **15**(1), 6596 (2014)
17. Rana, S., Gupta, S., Venkatesh, S.: Differentially-private random forest with high utility. In: ICDM, pp. 955–960. IEEE, Atlantic City (2015)
18. Gupta, S., Rana, S., Saha, B., Phung, D., Venkatesh, S.: A new transfer learning framework with application to model-agnostic multi-task learning. In: KAIS (2015)

Intelligent Recognition of Spontaneous Expression Using Motion Magnification of Spatio-temporal Data

B.M.S. Bahar Talukder[1], Brinta Chowdhury[1], Tamanna Howlader[2], and S.M. Mahbubur Rahman[1]([✉])

[1] Department of Electrical and Electronic Engineering, Bangladesh University of Engineering and Technology, Dhaka 1205, Bangladesh
sabquatfast@gmail.com, brinta.buet.09.eee@gmail.com, mahbubur@eee.buet.ac.bd
[2] Institute of Statistical Research and Training, University of Dhaka, Dhaka 1000, Bangladesh
tamanna@isrt.ac.bd

Abstract. The challenges of recognition of spontaneous expressions from spatio-temporal data include the characterization of subtle changes of facial textures, which in many cases occur for a very brief duration. In this context, the paper presents an intelligent approach for spontaneous expression recognition algorithm, wherein adaptive magnification of motion of spatio-temporal data is applied prior to the extraction of features of expression. The proposed magnification enhances the low-intensity facial activities without introducing notable artifacts for the high-intensity activities. The local binary patterns extracted from three-orthogonal planes of the Eulerian magnified spatio-temporal data are used as features of spontaneous expressions. The extracted features are classified using the well-known support vector machine classifier. Experiments are conducted on commonly-referred spatio-temporal databases such as the SMIC and MMI that have spontaneous expressions representing the micro- and meso-level facial activities, respectively. Experimental results reveal that the proposed approach of motion magnification prior to feature extraction significantly improves the detection and classification accuracy at the expense of acceptable robustness.

Keywords: Eulerian motion magnification · Expression features · Local binary patterns · Spatio-temporal data · Spontaneous expression

1 Introduction

In the recent years, the understanding of emotional state of humans from physiological traits has been gaining increasing research interest, especially in the area of security aware applications. This is mainly due to the fact that estimating the meaningful emotional state can be very useful for wide-deployment of interactions between humans and machines as well as for automatic analysis of

© Springer International Publishing Switzerland 2016
M. Chau et al. (Eds.): PAISI 2016, LNCS 9650, pp. 114–128, 2016.
DOI: 10.1007/978-3-319-31863-9_9

social behavior of people. The emotional state can be continuous in three independent spaces, namely, valance, arousal and dominance [1]. However, Ekman and Friesan [2] have shown that discrete emotional level can be categorized in six basic expressions, viz., Happy, Sad, Anger, Surprise, Fear, and Disgust, which are independent of cultures. The physiological traits that are used for recognition of expressions include the measurements of activities of faces via spatio-temporal imaging modalities [3], activities of tones via voice modalities [4], or even that of bio-signals, e.g., electromyography and electroencephalography [5]. Nevertheless, the spatio-temporal affective analysis has remained in the top-priority due to the fact that such modalities capture a significant amount of information in a non-intrusive manner.

Spatio-temporal expression classification methods can be broadly classified in three categories, namely, the model or geometric-based, holistic or appearance-based, and combination of these or hybrid approach [6]. In the geometric approach, certain key points on the face images are identified and the features of expressions are obtained from these fiducial points. For example, the inter distance among these points, the change of textures of the neighboring region of these points over the frames that is often referred to as the facial action units (FAUs) are considered as measures of facial activities. The main problem of the geometric approach lies in the selection of fiducial points, which often requires manual intervention even in controlled environments. Further, the accuracy of classification is highly sensitive to the localization of the fiducial points. At the same time, the geometric approach often requires computationally expensive algorithms such as the elastic bunch graph matching method to obtain the features of facial activities [7].

In the appearance-based approach, the entire facial region is considered for extracting features of expression instead of certain fiducial points. Conventional holistic approaches that have been used for extracting facial features include the principal component analysis (PCA), independent component analysis, Fisher discriminant analysis (FDA) with Asymmetry-Face, kernel PCA-FDA, non-negative matrix factorization, and mixture covariance analysis applied to the whole face. In the hybrid approach, features of expressions are extracted from the local neighboring regions of facial parts called patches, which are partitioned uniformly or selectively. Manifold learning of patches have been shown to be effective for classification of expressions in the cases of significant distortions of faces such as those due to occlusions [8]. The texture-based features of the uniform or selective patches of facial images that were used for expression classification include the scale invariant feature transform (SIFT), histogram of oriented gradients (HOGs), local binary patterns (LBPs), local directional number pattern, local directional patterns variance, multiscale Gaussian derivatives, Gabor, log-Gabor features, and the geometric orthogonal moments. Densely sampled facial features have also been used to determine the expressions in-the-wild [9]. The expressions in question are determined from the chosen features using well-known classification techniques such as the support vector machine (SVM), Bayesian dynamic network, and neural network. Suitable feature selection strategies including the

AdaBoost and bagging have also been applied to minimize the redundancy and maximize the relevancy to improve the classification performance.

In order to accommodate the depth information of facial parts, the 3D face images have also been used in addition to the 2D intensity image for expression analysis. For example, the SIFT, HOG, and LBP-based features have been fused together to classify expressions from 3D face images [10]. It is to be pointed out that due to the constrained settings of 3D imaging, the practical facial expression analysis still depends very much on the spatio-temporal information available in 2D image sequence. Ji and Idrissi [11] have recommended that LBPs obtained from three orthogonal planes (TOPs) of spatio-temporal data can effectively represent features of facial expressions. Recent surveys on spatio-temporal analysis for facial affective analysis can be found in [12,13].

In practice, there exist two major types of expressions, viz., the posed- and spontaneous-type, based on the exaggeration of facial activities available in spatio-temporal data. The involvement of professional guidance are involved while disposing the posed-type expressions, while touches of real-life such as subtle changes of facial textures exist in the spontaneous-type expressions. Studies reveal that spontaneous-type expressions are significantly different from posed-type expressions, and in general, the slightest change of facial activities in spatio-temporal data can be more important in the former than that in the latter. In our day-to-day life, the subtle facial activities may last for couple of seconds representing a meso-level spontaneous expression. On the other hand, in micro-level spontaneous expressions, when people try to conceal their emotions, the extraction of facial activities is even more challenging. This is mainly due to the fact that in such cases the facial changes occur in fraction of a second [14]. The recognition of micro-expressions serves as important clue for detecting lies that usually occur in high-stake situations when people know about serious consequences of lying or cheating. In the literature, there exists few number of research studies that focus on the automatic recognition of micro-expressions. For example, Shreve et al. [15] used strain patterns as a feature descriptor for spotting posed-type micro-expressions in spatio-temporal data. Polikovsky et al. [16] have used the HOGs as a descriptor for micro-expression recognition in posed scenario. An initial research carried out by Li et al. [14] has shown that the features obtained in terms of the LBP-TOP of spatio-temporal data can perform well for classifying the micro-expression even in the case of spontaneous scenario. In [17], class-specific pass bands of temporal filters have been prescribed for magnification of spatio-temporal data. This method detects the micro-expressions by recognizing that the low, mid-range, and high frequency temporal data correspond to three types of movements, viz., broad head, lip/brow, and eye/pupil. Discriminative learning of the bands of temporal filter is proposed for recognizing subtle facial expressions in [18]. Dynamics of depth information and dense motion field of faces while uttering certain vocabularies are also used to determine micro-expressions [19]. In [20], a set of arbitrarily chosen magnification factors for the region specific motion vectors of geometric face features was used to enhance the recognition performance of subtle expressions. In a recent report,

Li et al. [21] use peak contrast of feature difference to spot the micro-expressions in spatio-temporal data. In this algorithm, heuristically chosen ten discrete levels are applied to magnify the Eulerian motion of spatio-temporal data prior to the extraction of LBP or HOG features for recognizing the micro-expressions.

In this paper, we argue that instead of applying fixed-level of magnification of motion, certain adaptive magnification should be employed to enhance the subtle activities of spatio-temporal data for the purpose of recognition of spontaneous expressions. The objectives of adaptation of motion magnification is two-fold: (i) to amplify the low intensity motions in micro-expressions and (ii) to reduce excessive artifacts induced in the meso-level expressions due to magnification. In particular, we promote the use of simple yet effective mean-adaptive Eulerian motion magnification in conjunction with the LBP-TOP features for the representation of micro- or meso-level facial activities in the spontaneous expressions. Experimental results obtained from databases having micro-expressions, namely, spontaneous micro-expression (SMIC) and meso-level expressions, namely, M&M initiative (MMI) reveal that the proposed approach of adaptive motion magnification improves the performance of expression classification significantly.

The paper is organized as follows. Section 2 presents the proposed approach of adaptation for the Eulerian motion magnification. The feature extraction and classification of expressions are detailed in Sects. 3 and 4, respectively. The experimental results showing the significance of proposed approach for improving the classification performance is given in Sect. 5. Finally, Sect. 6 provides the conclusions.

2   Motion Magnification

Let $I(x, y; t)$ denote the spatial intensity at position (x, y) and time t in a spatio-temporal data of size (X, Y, T). Let the initial intensity in a given frame $I(x, y; 0)$ be denoted as $f(x, y)$. If the translational displacement called motion vector in time t is $\delta^{xy}(t) = \delta^x(t) + j\delta^y(t)$, where j is a complex operator, then $I(x, y; t) = f(x, y; \delta^{xy}(t))$. The Eulerian motion magnification of the data by a fixed-level α refers to synthesizing a signal given by [22]

$$I_m(x, y; t) = f(x, y; (1 + \alpha)\delta^{xy}(t)) \tag{1}$$

According to first-order Taylor series expansion around (x, y), the motion magnified signal can be approximated as

$$\tilde{I}_m(x, y; t) \approx f(x, y) + (1 + \alpha)\sqrt{\left(\delta^x(t)f_x\right)^2 + \left(\delta^y(t)f_y\right)^2} \tag{2}$$

where $f_x \equiv \partial f(x, y)/\partial x$ and $f_y \equiv \partial f(x, y)/\partial y$. In a general case, the selective-band temporal filter is used so that a good approximation of motion magnified signal can be attained [17]. Let $\delta_\omega^{xy}(t) = \delta_\omega^x(t) + j\delta_\omega^y(t)$ represent the different spectral components of $\delta^{xy}(t)$ in a continuous variable of temporal frequency ω. Let the frequency dependent motion magnification factor be α_ω. In such a case, the resultant motion magnified intensity is given by

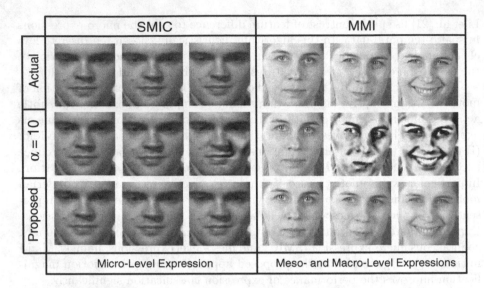

Fig. 1. Comparison of outputs of video frames due to motion magnification in micro-, meso-, and macro-level expressions. The top row of frames show the actual frames. The middle row shows the motion magnified frames when $\alpha = 10$. The bottom row shows the motion magnified frames as per the proposed approach.

$$\hat{I}_m(x, y; t) \approx f(x, y) + \int_\omega (1 + \alpha_\omega)\sqrt{\left(\delta_\omega^x(t)f_x\right)^2 + \left(\delta_\omega^y(t)f_y\right)^2}\,d\omega \qquad (3)$$

The first-order Taylor series expansion may introduce significant artifacts when spatial frequency is considerably high such as for noticeable changes in $f(x, y)$. In order to restrict the over magnification due to high spatial frequencies, the magnification factor α_ω is constrained according to the recommendation given in [22] as

$$(1 + \alpha_\omega)\delta_\omega^x(t) < \frac{\lambda_u}{8} \qquad (4)$$

$$(1 + \alpha_\omega)\delta_\omega^y(t) < \frac{\lambda_v}{8} \qquad (5)$$

where $\lambda_u = 2\pi/u_x$ and $\lambda_v = 2\pi/v_y$ are the wavelengths of spectral components of $f(x, y)$ that are expressed in terms of the continuous variables of spatial frequencies u_x and v_y, respectively. These restrictions may not work well for the meso-level spontaneous expressions and macro-level posed expressions, when the facial activities are non-trivial. Hence, an adaptive magnification of spatio-temporal data is required, in which the magnification level can be selected according to the overall motions available in the data. In particular, the adaptive magnification level can be inversely proportional to the mean of magnitude of displacement vectors available in the entire data given by

$$\hat{\alpha}_\omega = \eta XYT \left[\int_x \int_y \int_t \sqrt{\left(\delta_\omega^x(t)\right)^2 + \left(\delta_\omega^y(t)\right)^2} \, dxdydt \right]^{-1} \tag{6}$$

where η ($\eta \geq 1$) is a proportional constant. A value of η close to unity is preferable to avoid excessive magnification and resulting artifacts in the frames. In practice, any pixel-based motion estimation algorithm can be used to find the adaptive motion magnification factor $\hat{\alpha}_\omega$. In the proposed method, the block-matching and optical flow-based subpixel motion estimation technique is recommended due its fast implementation [23]. Figure 1 shows a typical comparison of the motion magnified frames of spatio-temporal data having micro-, meso-, and macro-level expressions that are available in the SMIC and MMI databases. It is seen from this figure that a heuristic choice of magnification factor such as $\alpha = 10$ can produce significant artifacts in spatial textures of the frames, which in fact severely affects the expression recognition performance. As expected, the artifacts caused due to improper magnification is increasingly pronounced in the case of micro-, meso-, and macro-level expressions. However, if the proposed adaptation of scaling of magnification is considered, then the artifacts are reduced significantly. It may be mentioned that the perceptual quality of motion magnification and reduced amount of artifacts provided by the proposed method appears to be even better than that shown in Fig. 1, when the entire frames of the spatio-temporal data are viewed rather than just a few number of frames.

3 Features for Expressions

In the proposed recognition algorithm, the features of expressions are extracted from the magnified spatio-temporal data using the LBP-TOP algorithm [24]. This descriptor is obtained by concatenating LBP on three orthogonal planes: XY, XT, and YT, and considering only the co-occurrence statistics in these three directions. Let us consider that a set of dynamic textures in terms of LBP of size $X_d \times Y_d \times T_d$ ($x_c \in \{1, 2, \cdots, X_d\}, y_c \in \{1, 2, \cdots, Y_d\}, t_c \in \{1, 2, \cdots, T_d\}$) are estimated by considering only the center part of the neighborhood in the magnified data. The histogram of the dynamic feature is estimated as

$$H_{i\ell} = \sum_{x,y,t} I\{\Phi_\ell^c(x, y, t) = i\} \quad i = 1, 2, \cdots, n_\ell \quad \ell = 1, 2, 3 \tag{7}$$

where $\Phi_\ell^c(x, y, t)$ expresses the uniform LBP code of central pixel (x_c, y_c, t_c), n_ℓ is the number of different labels produced by the LBP operator in the ℓth plane ($\ell = 1 : \text{XY}, 2 : \text{XT}, 3 : \text{YT}$) and

$$I\{A\} = \begin{cases} 1 & \text{if } A \text{ is true} \\ 0 & \text{if } A \text{ is false} \end{cases} \tag{8}$$

In order to get a coherent description, the histograms obtained are normalized as

$$N_{i\ell} = \frac{H_{i\ell}}{\sum_{k=1}^{n_\ell} H_{k\ell}} \quad i = 1, 2, \cdots, n_\ell \quad \ell = 1, 2, 3 \tag{9}$$

The labels from the XY-plane contain information about the appearance, whereas that in the XT- and YT-planes represent the co-occurrence statistics of motion in horizontal and vertical directions. The three histograms are concatenated to build a global description F that represents dynamic feature with the spatial and temporal characteristics of the data, which is often referred to as the LBP-TOP.

4 Feature Classification

In the proposed method, the kernel-based SVM is employed to classify the histograms of LBP-TOP by acknowledging that it is a well established statistical learning theory applied successfully in many classification tasks in computer vision. The kernel SVM implicitly maps the LBP-based features into a higher dimensional feature space to find a linear hyperplane, wherein the expressions can be categorized with a maximal margin. Given a training set of Γ labeled expressions $\Theta_{tr} = \{(F_\gamma, d_\gamma) | \gamma = 1, 2, \cdots, \Gamma\}$, where $F_\gamma \in \mathbb{Z}^{n_\ell}$ and $d_\gamma \in \{-1, 1\}$, the test feature F is classified using the decision function

$$\mathcal{D}(F) = \text{sign} \left(\sum_{\gamma=1}^{\Gamma} \beta_\gamma d_\gamma \Psi(F_\gamma, F) + b \right) \tag{10}$$

where β_γ are the Lagrange multipliers of a dual optimization problem that describe the separating hyperplane, $\Psi(F_\gamma, F)$ is a kernel function, and b is the weight of bias. The training samples F_γ with $\beta_\gamma > 0$ are called the support vectors. The SVM finds the separating hyperplane that maximizes the margin with respect to these support vectors. In order to map the LBP-based histogram into the higher dimensional feature space for classification, the most frequently used kernel functions such as the linear, polynomial, and radial basis function can be used.

Admitting that the SVM provides a binary decision, the multiclass decisions can be obtained by adopting the several two-class or one-against-rest problems. In the proposed method, one-against-rest problems are chosen, and hence ultimate expression class is obtained by Γ number of binary learners. With a view to select the parameters of the SVM, a grid-search on the hyper-parameters is used by adopting a cross-validation scheme. The parameter settings that produce the best cross-validation accuracy are used for obtaining the decision on the LBP-TOP feature under test.

5 Experimental Results

The experiments presented in this paper mainly focus on the effect of proposed motion magnification on the recognition of spontaneous expressions. The experiments are conducted on both the micro- and meso-level spontaneous expressions. In order to present representative results, only the findings of the recognition

Table 1. Results of Expression Classification Accuracy of SMIC-HS Dataset

Method	Class	Actual		Magnified	
		Overall (%)	Best (%)	Overall (%)	Best (%)
Detection	Micro	63.65±0.0016	69.05	65.34±0.1111	70.63
Recognition	Positive	37.46±0.0068	52.38	40.63±0.0175	57.14
	Negative	40.63±0.0233	71.43	44.13±0.0199	66.67
	Surprise	52.06±0.0199	76.19	53.97±0.0145	76.19

Table 2. Results of Expression Classification Accuracy of SMIC-VIS Dataset

Method	Class	Actual		Magnified	
		Overall (%)	Best (%)	Overall (%)	Best (%)
Detection	Micro	52.00±0.0039	61.67	54.88±0.0049	65.00
Recognition	Positive	59.33±0.0135	90.00	60.00±0.0243	100.00
	Negative	39.33±0.0107	50.00	49.33±0.0278	80.00
	Surprise	44.00±0.0126	60.00	48.00±0.0246	80.00

performance of micro-expressions of SMIC database and that of the meso-level expressions of the MMI database are presented. In this section, first we provide a brief description of the two datasets, which is followed by the experimental setup and the performance comparisons of expression recognition with and without magnification of motions of data.

5.1 Datasets

The SMIC is a spontaneous micro-expression database having 164 video clips, in which the involuntary emotions were induced by displaying audiovisual films and the required level of inhibition in expressing emotions were strictly maintained by imposing enough pressure to 16 participants [14]. The facial activities were captured using three types of cameras, viz., high speed (HS), normal visual (VIS) and near infra-red (NIR), all having pixel resolution of 640 × 480. The frame rate of HS camera is 100 fps and that of the rest two is 25 fps. The 71 video clips captured from the VIS and NIR cameras yield data similar to standard web cameras, including their limitations such as motion blurs. The micro-expressions in this dataset is classified into three classes (i) Surprise, (ii) Positive representing the emotion Happy and (iii) Negative representing any of the emotions Sad, Fear, or Disgust. The dataset did not elicit any micro-expressions for the emotion Anger. There is an extra class of video clips called non-micro which display no emotion though they have facial movements.

The MMI database has 197 video sequences of faces displaying mostly for the spontaneous-type facial expressions of one of the six basic emotions [25]. The video clips are collected from 75 subjects with a standard camera of frame rate 24

Table 3. Results of Expression Classification Accuracy of SMIC-NIR Dataset

Method	Class	Actual		Magnified	
		Overall (%)	Best (%)	Overall (%)	Best (%)
Detection	Micro	55.44±0.0039	66.67	57.00±0.0044	66.67
Recognition	Positive	58.67±0.0270	80.00	62.00±0.0246	90.00
	Negative	53.33±0.0152	70.00	57.33±0.0135	80.00
	Surprise	62.67±0.0421	90.00	67.33±0.0121	100.00

Table 4. Results of Expression Classification Accuracy of MMI Dataset

Method	Class	Actual		Magnified	
		Overall (%)	Best (%)	Overall (%)	Best (%)
Recognition	Happy	29.33±0.0292	50.00	31.33±0.0327	60.00
	Sad	32.00±0.0160	50.00	39.33±0.0235	70.00
	Surprise	48.67±0.0184	70.00	52.67±0.0121	80.00

fps and pixel resolution of 720×576. The sequences in the data corpus are fully annotated for the presence of single or multiple FAUs in the video. In order to generate a generic dataset of meso-level spontaneous expressions, we choose 25, 21, and 22 number of spatio-temporal clips from this database representing the Happy (i.e., Positive), Sad (i.e., Negative), and Surprise expressions, respectively. The face region is detected first using the Viola-Jones algorithm [26] and by using the coordinates of eye pair. Each spatio-temporal data representing the facial activities is cropped and resized to a spatial resolution of 180×240 using the bi-cubic interpolation.

5.2   Setups

To extract the LBP-TOP features from datasets, the frame lengths of clips of all expressions are normalized. In particular, the linear interpolation is used to normalize the frame length 20 for the clips of SMIC and 60 for the MMI datasets. The XY-plane of each of the clips is partitioned to 5×5, and the LBP-TOP features are calculated for all the partitions using the codes available in the website[1] maintained by the developers of LBP. These features are concatenated to obtain the ultimate features of expressions. In order to magnify the motion of a video, the parameter η is chosen to be close to unity. In the experiment, it is found that $\eta = 1.2$ works very well for both the SMIC and MMI datasets. The mean values of motions of spatio-temporal clips of datasets are obtained from the codes available in the website[2] of one of the authors of [23]. The motions are estimated using a block-size of 8×8 and a search limit of 10. The codes available

---

[1] http://www.cse.oulu.fi/CMV/Downloads/LBPMatlab.
[2] http://scholar.harvard.edu/stanleychan/software/.

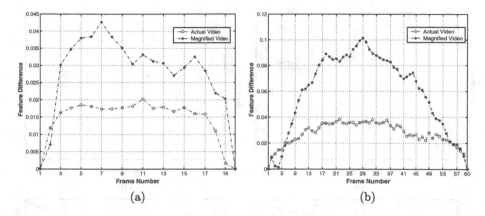

Fig. 2. Amplified feature intensities due to the proposed motion magnification of spatio-temporal clips of the datasets (a) SMIC and (b) MMI.

in the website[3] of one of the authors of [22] are employed for the Eulerian video magnification. The Laplacian pyramid with cutoff wavelength 16 is used for spatial filtering during the magnification. The temporal filtering is performed by using two infinite impulse response low-pass filters that have weights 0.4 and 0.05. The chrome attenuation factor is chosen as 0.1. These parameters are set as default in the codes for motion magnification.

The proposed expression recognition algorithm uses supervised learning technique by employing the kernel SVM. In particular, randomly chosen 50 % of the spatio-temporal clips for each of the expressions are treated as the training set, and the rest as the probe set. The expression recognition accuracies are estimated from the correctly classified clips in the probe set. The overall classification performance are reported in terms of the mean and standard deviation of the accuracies obtained from 15 independent randomly chosen training-probe sets. The hyper parameters of the kernel-based SVM classifier are estimated from the cross validation scheme applied only on the training sets. In the experiments, it has been found that the linear kernel function performs the best for the LBP-TOP-based expression classification.

5.3   Results

We first evaluate the changes of magnitude of the LBP-based features due to the proposed motion magnification. In particular, the LBP features are calculated for the XY-plane only, and the frame-by-frame feature differences are obtained from the last frame representing neutral expression of each of the spatio-temporal clips. Figure 2 shows typical frame-by-frame feature differences for the expression Surprise in the SMIC and MMI datasets for both the actual and motion magnified clips according to proposed approach. Since both the datasets have

---

[3] http://people.csail.mit.edu/mrub/evm/.

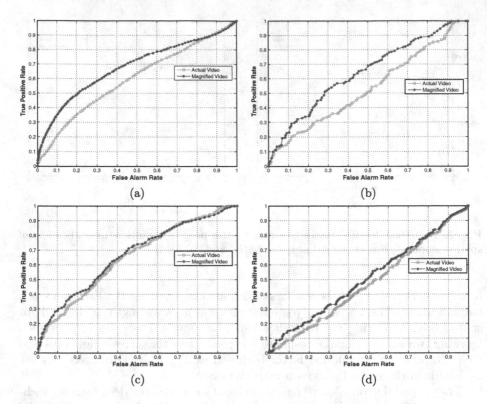

Fig. 3. Results concerning the ROC curves showing improved expression detection due to the proposed motion magnification of spatio-temporal clips. The curves are obtained for detection of (a) micro expressions from SMIC-HS clips, (b) Surprise expression from SMIC-VIS clips, (c) Positive expression from SMIC-NIR clips and (d) Happy expression from MMI clips.

the neutral expression at both ends of the clips, the feature distances slowly increase from the starting frame, reach a peak and then fall to the off-set. What is notable here in this figure is that due to the magnification of motion, the feature differences significantly increase from the actual ones. In other words, the proposed magnification of motion is capable of providing amplification of facial activities, which in turn is expected to provide improved performance for recognizing facial expressions.

Tables 1, 2 and 3 present comparisons of detection accuracy of micro expression and recognition accuracy of the Positive, Negative and Surprise expressions when the clips are magnified according to proposed approach and remain as actual in [14] for the three resolutions of SMIC dataset, viz., HS, VIS and NIR, respectively. It is seen from these tables that the detection accuracy of micro-level expressions is the highest for the HS camera, and it can be increased further by introducing the proposed motion magnification with a very slight compromise in the robustness. This is expected because, the motion magnification can slightly increase the chance of false detection, if the magnification factor is not scaled appropriately. In the case of recognition of types of expressions, the magnification of motion increases the accuracy by more than 3.9 % on average and shows relatively better performance improvement for the VIS and NIR clips. A negligible decrease in the robustness is seen for the HS and VIS clips, but noticeable improvement of robustness is observed for the low-resolution NIR clips especially for recognizing the Surprise expression. In all cases, the best detection or recognition accuracy of magnified video is never less than that without magnification. Table 4 presents the improvement of recognition accuracy for three expressions, namely, Happy, Sad, and Surprise in two scenarios when the spatio-temporal clips of MMI dataset are magnified or remain untouched. As can be seen from this table, even for the meso-level expressions, the improvement of average accuracy is more than 4.25 %, and the highest improvement is seen for the case of Sad expression. The robustness slightly decreases for the Happy and Sad expressions, and increases for the Surprise expression. But the improvements in the mean accuracy or that in the best accuracy due to the introduction of proposed motion magnification of spatio-temporal data actually surpass the slight sacrifice in the robustness.

The effect of magnification of spatio-temporal clips is also evaluated by estimating the receiver operating characteristics (ROC) curves for detection of micro-level expressions as well as for detecting a certain-type of expression from the pool of clips of a dataset. Figure 3 shows comparisons of ROC curves when the clips are magnified or remain as actual for typical scenarios in the SMIC and MMI datsets. In particular, Fig. 3(a) shows that the true positive rate for the detection of micro-expression in the HS clips of SMIC dataset significantly improves especially in the case of low-level false alarm rate, when the clips undergo the proposed motion magnification. Similar improvements are also observed for the recognition of Surprise and Positive expressions in the SMIC dataset as shown in Fig. 3(b) and (c), when the clips are captured by the VIS and NIR cameras, respectively. Due to the poor resolution, the improvement in the NIR clips due to motion magnification is observed to be marginal as compared to that in the HS and VIS clips of the SMIC dataset. Nevertheless, the motion magnification provides significantly higher values of true positive rate for a given false alarm rate for the MMI dataset, which is primarily dominated by the meso-level spontaneous expressions. A typical example of such improvement in detection accuracy of the expression Surprise is shown in Fig. 3(d). Thus, the proposed magnification invariably improves the detection or recognition performance both for the micro- or meso-level spontaneous expressions.

6   Conclusion

The most challenging aspect of detection and recognition of spontaneous expressions is the low-level of the facial activities in the captured spatio-temporal clips. Not only the duration of these facial activities is very brief in period, but also the trivial dynamics of the textures pose notable challenge to extract effective features of expressions. In order to overcome such problems, motion magnification prior to the extraction of features was recommended for detecting and classifying the micro-expressions. However, existing methods use heuristic choice to set the level of motion magnification. In such cases, artifacts are introduced for the facial activities, especially when there remains meso-level facial activities in the spontaneous expressions let alone the macro-level activities in posed expressions. Thus, the proposed paper has introduced a mean-adaptive level of motion magnification, so that small-scale dynamics of faces can be magnified without causing any significant artifacts. The features of the expressions can be extracted from the magnified clips with minimum error and thereby increasing the detection and classification accuracy.

In the proposed method, subpixel-based block-matching algorithm has been used for the motion estimation and the Eulerian technique for motion magnification. The LBP-TOP method has been adopted to extract the features for expression. It has been shown that the proposed magnification of spatio-temporal data enhances the feature intensities as a function of facial activities. The kernel SVM-based classification of features shows that the proposed method can significantly improve the mean accuracy of detection of micro-expression and that of classification of expressions for the SMIC dataset. Such improvements have been observed invariably for the three resolutions, namely, HS, VIS, and NIR of the spatio-temporal clips of SMIC dataset. The increased average classification has also been observed for the MMI dataset, which has meso-level activities representing the spontaneous expressions. The improvements of performance of detection of expressions have also been verified by constructing the ROC curves, which shows that the true positive rate for a given false alarm rate increases on average due to the motion magnification of the clips. We claim that such improvements have resulted due to the proposed adaptation of the level of motion magnification. The only challenge of the proposed magnification is the very tiny scale decay of the robustness of the detection and classification accuracies. But the results of overall classification accuracy, best accuracy, and the true positive rate clearly reveal that the use of proposed adaptation of motion magnification is worthy in detection or classification of spontaneous expressions. We expect that the proposed method can play a significant role for the next generation affective computing in the area of security and surveillance applications.

References

1. Wundt, W.M.: Grundzüge de physiologischen Psychologie. Engelman, Leipzig (1905)
2. Ekman, P., Friesen, W.V.: Constants across cultures in the face and emotion. J. Pers. Soc. Psychol. **17**(2), 124–129 (1971)
3. Wang, Z., Wang, S., Ji, Q.: Capturing complex spatio-temporal relations among facial muscles for facial expression recognition. In: Proceedings of the IEEE Conference on Computer Vision and Pattern Recognition, Portland, OR, pp. 3422–3429 (2013)

4. Bartlett, M.S., Littlewort, G.C., Frank, M.G., Lainscsek, C., Fasel, I.R., Movellan, J.R.: Automatic recognition of facial actions in spontaneous expressions. J. Multimedia **1**(6), 22–35 (2006)
5. Koelstra, S., Patras, I.: Fusion of facial expressions and EEG for implicit affective tagging. Image Vis. Comput. **31**(2), 164–174 (2013)
6. Pantic, M., Rothkrantz, L.J.M.: Automatic analysis of facial expressions: the state of the art. IEEE Trans. Pattern Anal. Mach. Intell. **22**(12), 1424–1445 (2000)
7. Ghimire, D., Lee, J., Li, Z.N., Jeong, S., Park, S.H., Choi, H.S.: Recognition of facial expressions based on tracking and selection of discriminative geometric features. Int. J. Multimedia Ubiquitous Eng. **10**(3), 35–44 (2015)
8. Ptucha, R., Tsagkatakis, G., Savakis, A.: Manifold based sparse representation for robust expression recognition without neutral subtraction. In: Proceedings of the IEEE International Conference on Computer Vision Workshops, Barcelona, Spain, pp. 2136–2143 (2011)
9. Kahou, S.E., Froumenty, P., Pal, C.: Facial expression analysis based on high dimensional binary features. In: Agapito, L., Bronstein, M.M., Rother, C. (eds.) ECCV 2014 Workshops. LNCS, vol. 8926, pp. 135–147. Springer, Heidelberg (2015)
10. Hu, Y., Zeng, Z., Yin, L., Wei, X., Zhou, X., Huang, T.S.: Multi-view facial expression recognition. In: Proceedings of the IEEE International Conference on Automatic Face & Gesture Recognition, Amsterdam, Netherlands, pp. 1–6 (2008)
11. Ji, Y., Idrissi, K.: Automatic facial expression recognition based on spatiotemporal descriptors. Pattern Recogn. Lett. **33**(10), 1373–1380 (2012)
12. Sariyanidi, E., Gunes, H., Cavallaro, A.: Automatic analysis of facial affect: a survey of registration, representation, and recognition. IEEE Trans. Pattern Anal. Mach. Intell. **37**(6), 1113–1133 (2015)
13. Wang, S., Ji, Q.: Video affective content analysis: a survey of state-of-the-art methods. IEEE Trans. Affect. Comput. **6**(4), 410–430 (2015)
14. Li, X., Pfister, T., Huang, X., Zhao, G., Pietikäinen, M.: A spontaneous micro-expression database: inducement, collection and baseline. In: Proceedings of the IEEE International Conference on Automatic Face and Gesture Recognition and Workshops, Shanghai, China, pp. 1–6 (2013)
15. Shreve, M., Godavarthy, S., Goldgof, D., Sarkar, S.: Macro- and micro-expression spotting in long videos using spatio-temporal strain. In: Proceedings of the IEEE International Conference on Automatic Face and Gesture Recognition and Workshops, Santa Barbara, pp. 51–56 (2011)
16. Polikovsky, S., Kameda, Y., Ohta, Y.: Facial micro-expressions recognition using high speed camera and 3D-gradient descriptor. In: Proceedings of the IET IEEE International Conference on Crime Detection and Prevention, London, UK, pp. 1–6 (2009)
17. Gogia, S., Liu, R.: Motion magnification of facial micro-expressions. Technical report 4, Massachusetts Institute of Technology (2014). http://runpeng.mit.edu/project#research
18. Park, S.Y., Lee, S.H., Ro, Y.M.: Subtle facial expression recognition using adaptive magnification of discriminative facial motion. In: Proceedings of the ACM IEEE International Conference on Multimedia, pp. 911–914 (2015)
19. Akagi, Y., Kawasaki, H.: A method of micro facial expression recognition based on dense facial motion data. In: Proceedings of the IEEE International Conference on Central European Computer Graphics, Visualization and Computer Vision, Plzen, Czech Republic, pp. 39–44 (2014)
20. Park, S., Kim, D.: Subtle facial expression recognition using motion magnification. Pattern Recogn. Lett. **30**(7), 708–716 (2009)

21. Li, X., Hong, X., Moilanen, A., Huang, X., Pfister, T., Zhao, G., Pietikäinen, M.: Reading hidden emotions: spontaneous micro-expression spotting and recognition. Technical report 1511.00423v1, Cornell University, arXiv e-prints (2015)
22. Wu, H.Y., Rubinstein, M., Shih, E., Guttag, J., Durand, F., Freeman, W.: Eulerian video magnification for revealing subtle changes in the world. ACM Trans. Graph. **31**(4), 1–8 (2012)
23. Chan, S.H., Vo, D.T., Nguyen, T.Q.: Subpixel motion estimation without interpolation. In: Proceedings of the IEEE International Conference on Acoustics Speech and Signal Processing, Dallas, TX, pp. 722–725 (2010)
24. Zhao, G., Pietikäinen, M.: Dynamic texture recognition using local binary patterns with an application to facial expressions. IEEE Trans. Pattern Anal. Mach. Intell. **29**(6), 915–928 (2007)
25. Pantic, M., Valstar, M., Rademaker, R., Maat, L.: Web-based database for facial expression analysis. In: Proceedings of the IEEE International Conference on Multimedia and Expo, Amsterdam, The Netherlands, pp. 1–5 (2005)
26. Viola, P., Jones, M.J.: Robust real-time face detection. Int. J. Comput. Vis. **57**(2), 137–154 (2004)

Cybersecurity
and Infrastructure Protection

k-NN Classification of Malware in HTTPS Traffic Using the Metric Space Approach

Jakub Lokoč[1], Jan Kohout[2], Přemysl Čech[1(✉)], Tomáš Skopal[1], and Tomáš Pevný[2]

[1] SIRET Research Group, Department of Software Engineering, Faculty of Mathematics and Physics, Charles University in Prague, Prague, Czech Republic
{okoc,cech,skopal}@ksi.mff.cuni.cz
[2] Department of Computer Science and Engineering, FEE, Czech Technical University in Prague, Cisco Systems, Inc., Cognitive Research Center in Prague, Prague, Czech Republic
{jkohout,tpevny}@cisco.com

Abstract. In this paper, we present detection of malware in HTTPS traffic using k-NN classification. We focus on the metric space approach for approximate k-NN searches over dataset of sparse high-dimensional descriptors of network traffic. We show the classification based on approximate k-NN search using metric index exhibits false positive rate reduced by an order of magnitude when compared to the state of the art method, while keeping the classification fast enough.

Keywords: Similarity search · k-NN classification · Intrusion detection

1   Introduction

Network Intrusion Detection Systems (NIDS) are presently an essential tool in detecting intrusions in computer networks, infected computers within, and other types of unwanted behaviour (e.g. exfiltration of company's sensitive data, using peer to peer networks, etc.). Traditionally, these systems have relied on the *signature matching* paradigm, which identifies known sequences of bytes (signatures) in packets unique for a particular virus, trojan, or other threat, which we further refer to as a *malware*. The advantage of signature matching is very low false alarm rate, however, malware nowadays implements plethora of evasion techniques such as polymorphism and encryption to randomize byte sequences rendering signature matching ineffective.

While randomizing byte sequences is relatively simple, randomizing behaviour is conceptually significantly more difficult problem. For example if the attacker wants to steal data from a computer, he needs to transfer them over the network. Similarly if he wants to know, if infected computers are still infected, they need to contact attacker's computer. Behaviour based NIDS detects such actions that are specific to malware activity, by using higher-level features of the traffic, such as the number of connections to different hosts during some

© Springer International Publishing Switzerland 2016
M. Chau et al. (Eds.): PAISI 2016, LNCS 9650, pp. 131–145, 2016.
DOI: 10.1007/978-3-319-31863-9_10

period of time, number of transferred bytes between two computers, etc. The drawback of behaviour based NIDS is higher false alarm rate, but the recall can be higher due to robustness with respect to simple randomization of byte sequences. The other advantage of relying on higher-level features is that they are exported by most network devices (switches, routers, HTTP proxies), which increases the visibility into the network (there are more collection points), and simplifies deployment as no adaptation of devices is needed.

This paper focuses on detecting secure HTTP (HTTPS) connections related to a malware activity, which is a pressing problem due to the generally growing volume of HTTPS traffic on the Internet (accelerated by Snowden's affair) and increasing adoption of HTTPS protocol by malware for its primary mean of communication. Information about HTTPS connections are effectively limited to the number of uploaded and downloaded bytes and a duration of the connection, which makes the classification of HTTPS connections particularly difficult. Nevertheless the prior art [13,14] has presented a statistical *fingerprint* of servers based on modelling joint distribution of properties of all connections to it. It has been demonstrated that they are usable for detecting malware and grouping servers with similar purpose.

In this paper we built upon these fingerprints, as features they rely on can be extracted from HTTPS connections. We show that (i) albeit the large dimension of fingerprints (14641), the problem of separating malware connections from legitimate is not linearly separable; (ii) a simple nearest neighbour based detector have order of magnitude better false alarm rate than the linear detector at the same recall; (iii) since we use large number of labeled data, the presented results estimate well the accuracy the representation of fingerprints can offer; (iv) we demonstrate that modern indexing structures allow to implement otherwise costly nearest-neighbour based detector efficiently, such that it become competitive to the linear classifier, particularly if its better false alarm rate is taken into the account.

2   Challenges in Detection of Malware Using HTTPS

The web proxy logs have become a widely used source of input data for the network intrusion detection systems because they provide relatively lightweight information about network that can be processed in high volumes to detect suspicious communication. However, the increasing usage of encrypted web communication via the HTTPS protocol hardens such detection or any traffic analysis at all. The creators of malware are aware of this fact which leads them to design the malware to use the encrypted communication as well. A commonly used way how to deal with HTTPS on web proxies is to intercept the HTTPS traffic on the proxy, decrypt it, log the information needed and encrypt it again. While the advantage is that the same detection techniques as in the case of HTTP traffic can be used, we see the main disadvantages in the computational burden on the proxy and the security concerns associated with the re-encryption of the private traffic. An alternative way is to develop new detection mechanisms that do not rely on features available only after the decryption. This is a challenging problem because the behavior of the network connections has to be reconstructed from very limited information. If the connection requests are not decrypted by the proxy, they are usually logged just as the so called *connect* requests [11], for which the only data available are the amounts of bytes transferred in both directions and the length of the time interval for which the connection was open. This rules out, for example, any detection methods that are based on features extracted from the visited URLs. On the other hand, if a detection method capable of working with such limited information is available, the design of the web proxy can be significantly simplified, its security improved and the privacy of the users is fully preserved. Therefore, developing such a method is a highly desirable task which motived our research presented in this paper.

3 Related Work

To the best of our knowledge, the prior art on using machine learning algorithm to classify HTTPS connections and detect malware is very limited. However there is some prior art using higher-level features as is the goal here. For example, Wright et al. [22] use a k-NN classifier with sizes and directions of TCP packets carrying encrypted traffic to identify application layer protocols. Contrary to our goal, their aim is not to distinguish between benign and malware's traffic, but to identify application protocols carried by the encrypted packets. Works of Crotti et al. [9] and Dusi et al. [10] use empirical estimates of probability distributions of packets' sizes and inter-arrival times in TCP flows for identification of application protocols (e.g. POP3, SMTP, HTTP). Despite that these do not aim directly to processing encrypted traffic, their representations could be applicable to it. The limiting factor of these approaches is that they need to know the order of each packet from the start of the TCP flow. Moreover, network sensors do not usually export information or statistics about individual packets within the connection. Many works can be found in the field of behavioral detection of malware that use non-encrypted HTTP traffic. Although they do not apply content inspection, they still employ features that can not be extracted from HTTPS traffic, for example lengths of URLs or types of HTTP methods (e.g., GET or POST) — for examples of such works, see [19,20] or [16].

In the work of Kohout and Pevny [13], servers are represented by a joint histogram of tuples $(r_{up}, r_{down}, r_{td}, r_{ti})$ with 11^4 bins, where r_{up} is the number of bytes sent from the client to the server, r_{down} is the number of bytes received by the client from the server, r_{td} is the duration of the connection (in milliseconds), and r_{ti} is the time in seconds elapsed between start of the current and previous request of the same client. Each tuple describes one connection to the server, and the dynamic range of values is decreased by taking $log(1 + x)$ before creating the joint the histogram. The biggest advantage of this representation is that the information is typically exported by most network devices supporting IPFIX [8] and Netflow [6] formats, web proxies, and they are available for encrypted (HTTPS) connections. With respect to this, these fingerprints are used in our work as basic features. Their dimension is $11^4 = 14641$ for each server.

In the introduction, it is mentioned that the biggest advantage of signature based techniques is low false alarm rate. This property is crucial for practical deployment, since the problem is unbalanced with the prevalence of legitimate traffic. For example, in our experimental dataset there are 160–220 thousands of benign servers and only 600–1800 examples of those that were related to malware activity (depending on a particular testing set). Thus high false alarm rate floods the network operator with meaningless alarms, which renders the system useless. Due to this imbalance in costs of error, the performance of each classifier is measured by the false alarm rate at 50 % recall on malware [21] (further called FP-50). Formally, the value of the FP-50 error e is defined as follows:

$$e = \frac{1}{|\mathcal{I}^-|} \sum_{i \in \mathcal{I}^-} \mathrm{I}\left[f(x_i) > \mathrm{median}\{f(x_j) | j \in \mathcal{I}^+\}\right], \tag{1}$$

where $\mathcal{I}^-$ are indexes of the negative (benign) training samples, $\mathcal{I}^+$ are indexes of the expositive training samples (malware), $f(x)$ is the output of the classifier (classification score) for the sample x and I is an indicator function. The rationale behind the measure is that we are willing to miss 50 % of malware for the benefit of having extremely low false positive rate. Moreover, thanks to the using of the 2-quantile of the false negatives (which reflects the demanded 50 % recall), the median in (1) can by replaced with mean for purposes of the FP-50 minimisation. This makes the optimization computationally tractable. In [21], the exponential Chebyshev Minimizer (ECM) was presented as a suitable linear classifier optimizing the FP-50 measure. This classifier is used also in the experimental section as a baseline classification technique. Denoting $x \in \mathbb{S} \subset \mathbb{R}^d$ the training set of samples, ECM solves the following optimization problem

$$\arg\min_{w \in \mathbb{R}^d} \frac{\lambda}{2} \|w\|^2 + \frac{1}{|\mathcal{I}^-|} \sum_{i \in \mathcal{I}^-} \log\left(1 + \exp(w^{\mathrm{T}}(x_i - \mu^+))\right), \tag{2}$$

where μ^+ is the mean of the positive samples: $\mu^+ = \frac{1}{|\mathcal{I}^+|} \sum_{i \in \mathcal{I}^+} x_i$. λ is a regularization parameter that needs to be set in prior of the training. Under reasonable assumptions on symmetry of the distribution of positive samples, the mean well approximates the median and ECM optimizes FP-50 for the class of linear functions. The optimization problem (2) is solved using L-BFGS method.

This paper investigates also a simple nearest-neighbour based classifier considering the Euclidean norm L_2 as a distance between two samples. The classification rule for a test sample x is based on its k nearest neighbours from a given training set $\mathbb{S}$. The k-NN query is defined for $k \in \mathbb{N}^+$, $x \in \mathbb{R}^d$ and $\mathbb{S} \subset \mathbb{R}^d$ as:

$$kNN(x) = \{\mathbb{X} \subset \mathbb{S}; |\mathbb{X}| = k \wedge \forall y \in \mathbb{X}, z \in \mathbb{S} - \mathbb{X} : L_2(x, y) \leq L_2(x, z)\}.$$

The k-NN classifier used in this paper assigns two values to each test sample x. The first value v_1 is the number of malicious objects in $kNN(x)$ and the second value v_2 is the sum of distances to the malicious objects in $kNN(x)$. The test samples are then sorted in the multi-column manner, where the samples are first sorted by v_1 in descending order and then by v_2 in the ascending order. This sorting is the input for the FP-50 measure.

Although being currently out of fashion the advantage of k-NN classifier is its universal consistency, which means that it converges to the optimum classifier in a Bayesian sense [2]. This property enables estimation of the accuracy that can be achieved using the representation of servers proposed in [13], provided sufficient number of well labelled samples is available (more on this in Sect. 5). As the nearest neighbour classifier can be extremely slow, the next section is devoted to presentation of indexing structures, which significantly improve the classification time. Since the data considered in this work are high-dimensional vectors, an approximate k-NN search strategy has to be employed by indexes to provide practical query processing times. The approximate search, however, affects classification, because the set of nearest neighbors just approximates the real nearest neighbors. Nevertheless, in the experiments we show that the approximation of the real nearest neighbors has negligible impact on the accuracy.

4   Efficient k-NN Search Using Metric Indexing

To perform the kNN classification fast, a database index is needed that provides exact and/or approximate kNN search over large set of high-dimensional sparse descriptors of the traffic. During last decades, there have been investigated many approaches to search efficiently in huge collections of (sparse) high-dimensional vectors. The approaches use various techniques to reduce the negative effect of high dimensionality of the data. For example, the dimension reduction techniques that try to find new low-dimensional representations of the original vectors preserving distances, or, the locality sensitive hashing [12] that tries to map close vectors to the same buckets with high probability. Within the vast portfolio of database techniques implementing the two principles, in this paper we focus on the metric index (M-Index [17]) that represents a metric variant of locality sensitive hashing [18]. The M-Index is suitable for the mentioned task as it represents efficient yet extensible solution for fast similarity searches.

4.1   Metric Indexing in a Nutshell

The fundamental trick of metric indexes lies in using lower bounds that can be used to filter out irrelevant object from the search cheaply (i.e., without the need of actual distance computations regarded as computationally expensive).

A *metric space* $(\mathbb{U}, \delta)$ consists of a descriptor domain $\mathbb{U}$ and a distance function δ which has to satisfy the metric postulates of *identity*, *non-negativity*, *symmetry*, and *triangle inequality*, defined $\forall x, y, z \in \mathbb{U}$ as:

$$\delta(x, y) = 0 \quad \Leftrightarrow \quad x = y \qquad \text{identity}$$
$$\delta(x, y) \geq 0 \qquad \text{non-negativity}$$
$$\delta(x, y) = \delta(y, x) \qquad \text{symmetry}$$
$$\delta(x, y) + \delta(y, z) \geq \delta(x, z) \qquad \text{triangle inequality}$$

In this way, metric spaces allow domain experts to model their notion of content-based similarity by an appropriate descriptor representation and distance function serving as similarity measure[1]. At the same time, this approach allows to design index structures, so-called *metric indexes* [1,3–5,17,23] for efficient query processing of content-based similarity queries in a database $\mathbb{S} \subset \mathbb{U}$. These methods rely on the distance function δ only, i.e., they do not necessarily know the structure of the descriptor representation of the objects.

Metric indexes organize database objects (descriptors) $o_i \in \mathbb{S}$ by grouping them based on their distances, with the aim of minimizing not only traditional database cost like I/O but also the number of costly distance function evaluations. For this purpose, nearly all metric indexes apply some form of filtering based on cheaply computed lower bounds. These bounds are constructed based on the fact that exact pivot–object distances are pre-computed, where a pivot is either a static or a dynamic reference object selected from the database.

We illustrate this fundamental principle in Fig. 1a where we depict the query object $q \in \mathbb{U}$, some pivot object $p \in \mathbb{S}$, and a database object $o \in \mathbb{S}$ in some metric space. Given a range query (q, r), we wish to estimate the distance $\delta(q, o)$ by making use of $\delta(q, p)$ and $\delta(o, p)$, with the latter already stored in the metric index. Because of the triangle inequality, we can safely filter object o without needing to compute the (costly) distance $\delta(q, o)$ if the triangular lower bound $\delta_T(q, o) = |\delta(q, p) - \delta(o, p)|$ is greater than the query radius r.

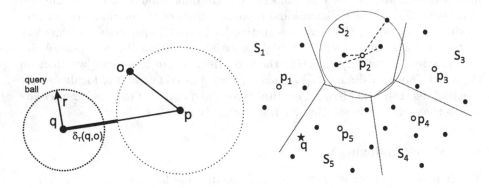

Fig. 1. (a) The lowerbounding principle, (b) The index.

4.2   Metric Index for kNN Search in High-Dimensional Sparse Data

There have been developed many metric indexes varying in the application purpose, however, in this paper we introduce a simple main-memory variant of the

---

[1] In our case, the descriptors are high-dimensional sparse vectors representing network traffic and the distance function is the Euclidean distance.

state-of-the-art structure M-Index that fits the requirements of network traffic kNN classification (high-dimensional data, cheap distance, many queries). Inspired by GNAT [1], M-index [17] and M-tree [5], we assume a set of p pivots p_i selected from the dataset $\mathbb{S}$, each representing a partition S_i in the metric space. The rest of the objects in the dataset are partitioned such that each object is assigned to a partition of its closest pivot. In such a way we obtain Voronoi partitioning (complete and disjoint), see Fig. 1b. Moreover, for each partition we store the distance r_i between the pivot and the furthest objects within the partition (partition radius), i.e., $r_i = \delta(p_i, o), o \in \mathbb{S}_i, \forall o_j \in \mathbb{S}_i : \delta(p_i, o_j) \leq \delta(p_i, o)$, see the dotted circle in the figure for r_2. Moreover, for each partition we also store the distances from the partition objects to the pivot p_i, i.e., $\langle \delta(p_i, o_j) \rangle_{o_j \in \mathbb{S}_i}$, see the dashed lines in the figure for $\delta(p_2, o_j)$.

The structure corresponds to one level GNAT or M-index, i.e., without additional (repetitive) Voronoi partitioning. The structure stores radius of each cluster and distances from partition objects to corresponding partition pivots as utilized by the M-Tree [5] (so called *distances to parent*). Note that also the M-Index considers the distances from partition objects to partition pivots to construct a key value that is stored in the B-Tree structure and used later for efficient searching. Since the distance function is not expensive, the structure stores information just for the most efficient filtering rules (ball-region filtering, parent filtering [23]), skipping less efficient filtering rules presented for M-Index (or GNAT) and expensive distance measures. In the following section, the approximate kNN search principles are detailed given the presented structure.

Approximate kNN Search. Given a kNN query (q, k), the index is used to process the query in a simple way. The partition with closest pivot to q is searched, then the partition with the second closest pivot is searched, and so on. During the search the closest k objects from the partitions searched so far to q are maintained as the kNN candidates. In order to speedup the search, there are several optimizations used.

First, when a partition is to be searched, its radius is checked whether the partition ball (p_i, r_i) overlaps the query ball or not. The query is formed by a ball centered in q having the radius to the actual kth nearest neighbour candidate. If the query ball and the partition ball do not overlap, the partition is skipped.

Second, whenever an object o from a partition is to be checked whether it is contained within the actual query ball or not, prior to computing the distance $\delta(q, o)$ the triangular lower bound is computed (as depicted in Fig. 1a). If the lower bound is greater than the actual query radius, the object cannot become a kNN candidate so the object is discarded without computing $\delta(q, o)$.

Third, as will be shown in experiments, it is not necessary to perform exact kNN search that could lead to exhaustive search in many of the partitions. In order to speedup the search even more, the kNN search could be stopped after a predefined number of objects is inspected (e.g., 4 % of all objects in the dataset). For the kNN pseudocode see Algorithm 1.

Algorithm 1. kNN(q, k, S, maxObjectsInspected)

1: compute distance to pivots $d(q, p[i])$ and sort partitions S[i]
2: initialize kNN candidate set Can using pivots $p[i]$
3: get actual query radius r
4: count = 0;
5: // for each partition check its objects
6: **foreach** S[i] **in** S
7:     **if** count > maxObjectsInspected **then break**
8:     // query-partition overlap check
9:     **if** $d(q, p[i]) > r[i] + r$ **then continue**
10:     **foreach** o[j] **in** S[i]
11:         // lower bound filter
12:         **if** $|d(q, p[i]) - d(o[j], p[i])| > r$ **then continue**
13:         distance = d(q, o[j])
14:         **if** distance $\geq r$ **then continue**
15:         update Can by o[j]
16:         $r = d(q, Can[k])$
17:         count++
18: **return** Can

5   Experiments

Below experiments reflect the problem to solve, which is the identification of servers contacted by malware by observing HTTPS connections to them. The first part focuses on description of the dataset and its labeling, which is an important aspect in the network intrusion detection research. Then it moves to the comparison of linear classifier to the nearest-neighbour based one, and finishes with a discussion of its scalability.

5.1   Dataset

The experimental dataset contains logs of HTTPS connections observed during the period of one day (24 h) in November 2014 from 500 major international companies[2] collected using Cisco's cloud web security solution [7]. Each log of HTTPS connection contains source and destination IP address, number of sent and received bytes, duration of the connection, and timestamp indicating when the connection has started. Besides this, some logs contain SHA hash of the process that has initiated the HTTPS connection. These logs with hashes are used in the experiments here, since matching them with a database of malware hashes[3] provides the precious labels (malware/benign). We emphasize here that the process hash is usually missing in logs, since network devices like routers, switches, and proxies do not have any means to compute them. Also, the availability of the hash is independent to the type of the connection, therefore errors measured on this set is a good estimate of what can be expected in real world.

---

[2] The exact cannot be published due to non-disclosure agreements.
[3] Specifically, the hash was considered to be malicious if the corresponding process was detected by at least 20 anti-viruses used by virustotal.com service.

In total, there was 145 822 799 connections to 475 605 unique servers with the total volume of transferred data 10 082 GB. As already mentioned above, a request is deemed to be related to a malware activity if hash of its parent process was in Virustotal's[4] database of malware hashes. Similarly a domain or destination IP address is considered to be malware if there was at least one malware connection to it.

Since the subject of the interest are servers, a server's feature vector (a descriptor or fingerprint) is formed from a joint histogram of four-tuple $r = (r_{up}, r_{down}, r_{td}, r_{ti})$[5] of all HTTPS connections to it, as described in Sect. 3.

The employed dataset has specific properties. The 4-dimensional vectors used to create the joint histograms form a strongly uneven distribution in the corresponding 4-dimensional space, making some parts of the space (and thus corresponding bins of the fingerprints) dominant. Based on these observations, we have also calculated joint histograms considering inverse document frequency (idf) weighting. However, according to our experiments (see Fig. 2) the idf weighting just deteriorates the performance of the k-NN classifier.

5.2   Test Settings

The main error measure used for the comparison is the FP-50, which is the probability of false alarm at 50 % recall. Note that FP-50 is evaluated after all query objects with assigned labels are sorted using ranking obtained from the employed classifier. The rationale behind the FP-50 measure is that low false alarm rate is extremely important, hence the measure focuses on it. Moreover, since malware usually uses more than one server, 50 % recall is perfectly reasonable.

FP-50 and other quantities were estimated using six-fold cross-validation, where each fold contained all HTTPS connections observed during continuous four hours. Domain descriptors were calculated separately from connections in the training folds and in the testing fold. The exact number of domains (samples) is shown in Table 1.

The linear ECM classifier [21] has only one parameter, which is the regularization constant λ (see Eq. 2). Although the parameter has an influence on the accuracy, for a small values its effect on it is limited, which is caused by the large number of training samples. In all experiments below its value was set to 10^{-8}. The basic nearest neighbor classifier has also one tunable parameter, which is the number of nearest neighbors. The level of approximation of the k-NN search is another parameter affecting precision. The effect of both parameters on FP-50 measure is studied below in Fig. 4. The metric index uses 1000 randomly selected pivots.

---

[4] virustotal.com.

[5] r_{up} is the number of bytes sent from the client to the server, r_{down} is the number of bytes received by the client from the server, r_{td} is the duration of the connection (in milliseconds), and r_{ti} is the time in seconds elapsed between start of the current and previous request of the same client.

Table 1. The number of benign/malware/total servers (samples) in each fold in the cross-validation.

#domains	fold					
	1	2	3	4	5	6
in training set	444540	441357	441073	424555	428635	440690
in testing set	161908	165132	161352	224004	222294	192691
benign	160911	164276	160741	222605	220549	191204
malware	997	856	611	1399	1745	1487

5.3   Performance of the Classifiers

FP-50 together with training and classification times is shown in Table 2. The results reveal that the linear classifier (ECM) is outperformed in FP-50 by almost order of magnitude by the 4-nearest neighbor classifier (without idf weighting), which means that the problem is not linearly separable despite the high dimension. The presented times[6] show that efficient indexing techniques are necessary for k-NN classifier to reach practical times.

Table 2. FP-50 estimated by the cross-validation of compared classifiers, together with their average training and classification times.

classifier	FP-50	time of	
		training	classification
ECM	13.23 %	56mins	0.3s
exact 4-NN no index	2.015 %	0s	63mins
exact 4-NN	2.015 %	44s	17mins
exact 4-NN with idf	2.247 %	44s	23mins
approx. 4-NN (4 % DB)	2.017 %	44s	3mins

Figure 2 shows FP-50 measure of exact search based k-NN classifier with respect to k and two types of fingerprints — with and without tf-idf weighting. The advantage of a small k in the k-NN classifier can be explained by a non-linearity of the problem and very low number of malware domains, which amounts only of 0.6 % of the total number (see Table 1). Indeed, a t-sne plot [15] in Fig. 3 reveals that malware domains do not form a tight clusters, but they are scattered in the space and surrounded by legitimate ones. This means that increasing k increases the proportion of benign domains in the neighbourhood yielding into incorrect classification.

---

[6] The experiments have run on 64-bit Windows Server 2008 R2 Standard with Intel Xeon CPU X5660, 2.8 GHz, 12 cores supporting hyper-threading. The training of the ECM classifier has run on a virtual machine (VMWare) using 8 cores CPU 2.2 GHz and 132 GB RAM. Matlab library MinFunc has been used.

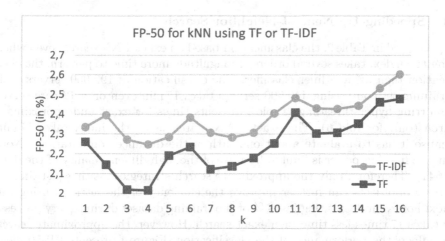

Fig. 2. FP-50 measure for exact kNN classifier and different types of descriptors.

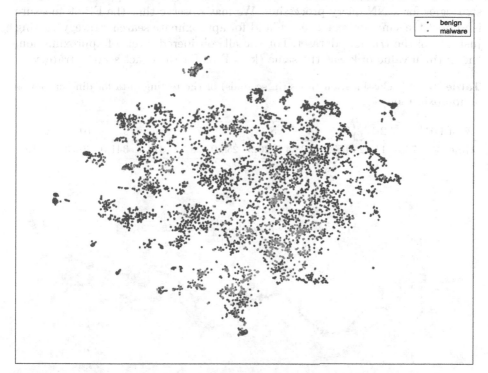

Fig. 3. A t-sne visualization of the space of servers' fingerprints, blue/red dots represent benign/malware fingerprints. The plot was created from uniformly sampled 5000 instances of benign fingerprints and 1000 instances of fingerprints of servers contacted by malware from the fold 6 of the cross-validation. This figure demonstrates that the malware's behavior does not follow a simple pattern common to all servers. However, as many of the servers contacted by malware form small clusters, the k-NN classifier which leverages the local similarities has a good chance to succeed (Color figure online).

5.4   Speeding-Up Nearest-Neighbor Search

As presented in Table 2, the classification based on exact 4-NN search, even when using the index, takes several orders of magnitude more time to perform the classification than using a liner classifier. The classification of 192.000 servers with a training set comprising 440.000 servers takes 17 min even on a 12-core server supporting hyper-threading. Table 3 presents times for exact and approximate search (only for fold 6). The exact 4-NN search using the index takes 17 min because it has to evaluate about one fifth of the distance computations. Note that the filtering power is limited because of the high-dimensionality of the data (14641). Therefore, only the approximate search strategies visiting just limited number of objects can further improve the efficiency of the retrieval using the index. For example, visiting just 1 % of the training dataset during query processing takes 17 times less time then exact search. However, the approximate search also affects the performance of the classification. Figure 4 presents FP-50 measure for the k-NN classification variants considering faster approximate search strategies for k-NN query processing. We may observe that the FP-50 measure is almost the same for exact search and for approximate search strategy visiting just 4 % of the training dataset. For the all considered levels of approximation, the optimal value of k was the same (k = 4) as for the exact search strategy.

Table 3. 4-NN classification times (in seconds) of the testing data for different levels of approximation.

% of DB	1	2	3	4	5	6	7	8	9	10	100
time	73,1	118,2	157,1	193,4	226,9	258,0	287,2	314,7	341,0	367,3	1021,6

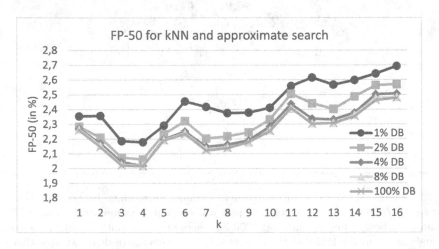

Fig. 4. FP-50 measure for the k-NN classifier and different levels of approximation.

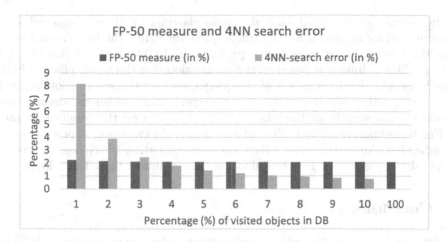

Fig. 5. FP-50 measure and approximation error for 4-NN classifier and different levels of approximation.

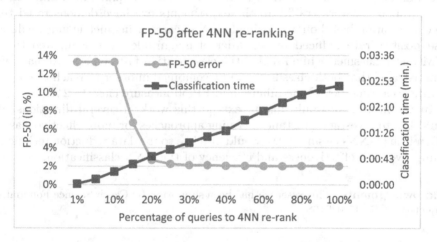

Fig. 6. FP-50 measure after re-ranking of ECM-based classification using k-NN classifier. Note that the time for classification corresponds to the secondary axis.

Figure 5 illustrates the approximation error[7] with decreasing number of visited objects during 4-NN query processing. Albeit the approximation error significantly increases for lower percentage of visited objects, the value of the FP-50 measure changes only slightly. Such advantageous trade-off is promising for our future research focusing on even faster approximate search algorithms for k-NN classification of HTTPS traffic data.

---

[7] For a given query, the approximation error is computed as a normed overlap distance between the query result returned by approximate k-NN search and the correct result returned by exact k-NN search.

In the last set of experiments, the linear classifier ECM was employed as an efficient ranking approach for the whole query set, while the investigated k-NN classifier was used to re-rank only a prefix of the ECM-based ordering of query objects. This technique is based on the assumption that the highly efficient ECM classifier could identify a subset of the query objects with low precision but high recall (i.e., with more than 50 % objects representing malicious servers) thus avoiding k-NN search for the whole query set. Based on the results presented in Fig. 6, re-ranking of just 30 % of the query objects using expensive k-NN classifier has almost the same effectiveness as evaluating k-NN classifier for all query objects, while the FP-50 evaluation time is almost three times lower.

6   Conclusions

In this paper, we have presented a technique for detection of malware in HTTPS traffic using k-NN classification. We have presented the efficiency of metric indexing for approximate k-NN search over dataset of sparse high-dimensional descriptors of network traffic. In the experiments, we have demonstrated that the classification based on approximate k-NN search using metric index exhibits false positive rate reduced by an order of magnitude when compared to the ECM linear classifier, while keeping the classification fast enough. We have also demonstrated that both classifiers can be combined in order to reach the overall classification time below one minute and FP-50 measure close to 2 %.

In the future, we would like to extend this work in several directions. We would like to investigate distance learning approaches for more effective classification using k-NN classifiers. We would also like to try data reduction techniques to improve both effectiveness and efficiency of the k-NN classification.

Acknowledgments. This research has been supported by Czech Science Foundation project (GAČR) 15-08916S.

References

1. Brin, S.: Near neighbor search in large metric spaces. In: Proceedings of 21th International Conference on Very Large Data Bases, VLDB 1995, 11–15 September 1995, Zurich, Switzerland, pp. 574–584 (1995). http://www.vldb.org/conf/1995/P574.PDF
2. Chaudhuri, K., Dasgupta, S.: Rates of convergence for nearest neighbor classification. In: Advances in Neural Information Processing Systems (2014)
3. Chávez, E., Navarro, G.: A compact space decomposition for effective metric indexing. Pattern Recogn. Lett. **26**(9), 1363–1376 (2005). http://dx.doi.org/10.1016/j.patrec.2004.11.014
4. Chávez, E., Navarro, G., Baeza-Yates, R., Marroquín, J.L.: Searching in metric spaces. ACM Comput. Surv. **33**(3), 273–321 (2001)
5. Ciaccia, P., Patella, M., Zezula, P.: M-tree: an efficient access method for similarity search in metric spaces. In: VLDB 1997, pp. 426–435 (1997)

6. Cisco: Cisco   IOS   NetFlow.   http://www.cisco.com/c/en/us/products/
   ios-nx-os-software/ios-netflow/index.html
7. Cisco: Cloud Web Security (CWS). http://www.cisco.com/c/en/us/products/
   security/cloud-web-security/index.html
8. Claise, B., Trammell, B., Aitken, P.: Specification of the IP Flow Information
   Export (IPFIX) Protocol for the Exchange of Flow Information (2013). https://
   tools.ietf.org/html/rfc7011
9. Crotti, M., Dusi, M., Gringoli, F., Salgarelli, L.: Traffic classification through simple
   statistical fingerprinting. SIGCOMM Comput. Commun. Rev. **37**, 5–16 (2007)
10. Dusi, M., Crotti, M., Gringoli, F., Salgarelli, L.: Tunnel hunter: detecting
    application-layer tunnels with statistical fingerprinting. Comput. Netw. **53**, 81–
    97 (2009)
11. Fielding, R., Gettys, J., Mogul, J., Frystyk, H., Masinter, L., Leach, P., Berners-
    Lee, T.: Hypertext Transfer Protocol – HTTP/1.1. https://tools.ietf.org/html/
    rfc2616
12. Gionis, A., Indyk, P., Motwani, R.: Similarity search in high dimensions via hash-
    ing. In: Proceedings of the 25th International Conference on Very Large Data
    Bases, VLDB 1999, pp. 518–529. Morgan Kaufmann Publishers Inc., San Fran-
    cisco (1999). http://dl.acm.org/citation.cfm?id=645925.671516
13. Kohout, J., Pevny, T.: Automatic discovery of web servers hosting similar applica-
    tions. In: 2015 IFIP/IEEE International Symposium on Integrated Network Man-
    agement (IM) (2015)
14. Kohout, J., Pevny, T.: Unsupervised detection of malware in persistent web traffic.
    In: 2015 IEEE International Conference on Acoustics, Speech and Signal Processing
    (ICASSP) (2015)
15. van der Maaten, L., Hinton, G.E.: Visualizing high-dimensional data using t-SNE.
    J. Mach. Learn. Res. **9**, 2579–2605 (2008)
16. Nelms, T., Perdisci, R., Ahamad, M.: Execscent: mining for new c&c domains in
    live networks with adaptive control protocol templates. In: Proceedings of the 22nd
    USENIX Conference on Security (2013)
17. Novak, D., Batko, M., Zezula, P.: Metric index: an efficient and scalable solution
    for precise and approximate similarity search. Inf. Syst. **36**(4), 721–733 (2011)
18. Novak, D., Kyselak, M., Zezula, P.: On locality-sensitive indexing in generic metric
    spaces. In: Proceedings of the Third International Conference on SImilarity Search
    and APplications, SISAP 2010, pp. 59–66. ACM, New York (2010). http://doi.
    acm.org/10.1145/1862344.1862354
19. Perdisci, R., Ariu, D., Giacinto, G.: Scalable fine-grained behavioral clustering of
    HTTP-based malware. Comput. Netw. **57**, 487–500 (2013)
20. Perdisci, R., Lee, W., Feamster, N.: Behavioral clustering of HTTP-based malware
    and signature generation using malicious network traces. In: Proceedings of the 7th
    USENIX Conference on Networked Systems Design and Implementation (2010)
21. Pevny, T., Ker, A.D.: Towards dependable steganalysis. In: IS&T/SPIE Electronic
    Imaging (2015)
22. Wright, C., Monrose, F., Masson, G.M.: On inferring application protocol behaviors
    in encrypted network traffic. J. Mach. Learn. Res. **7**, 2745–2769 (2006)
23. Zezula, P., Amato, G., Dohnal, V., Batko, M.: Similarity Search: The Metric Space
    Approach. Springer, New York (2005)

A Syntactic Approach for Detecting Viral Polymorphic Malware Variants

Vijay Naidu$^{(\boxtimes)}$ and Ajit Narayanan

School of Computer and Mathematical Sciences,
Auckland University of Technology, Auckland, New Zealand
{vijay.naidu,ajit.narayanan}@aut.ac.nz

Abstract. Polymorphic malware is currently difficult to identify. Such malware is able to mutate into functionally equivalent variants of themselves. Modern detection techniques are not adequate against this rapidly-mutating polymorphic malware. The age-old approach of signature-based detection is the only one that has the highest detection rate in real time and is used by almost all antivirus software products. The process of current signature extraction has so far been by manual evaluation. Even the most advanced malware detection process which employs heuristic-based approaches requires progressive evaluation and modification by humans to keep up with new malware variants. The aim of the research reported here is to investigate efficient and effective techniques of string matching algorithm for the automatic identification of some or all new polymorphic malware. We demonstrate how our proposed syntactic-based approach using the well-known string matching Smith-Waterman algorithm can successfully detect the known polymorphic variants of JS.Cassandra virus. Our string-matching approach may revolutionize our understanding of polymorphic variant generation and may lead to a new phase of syntactic-based anti-viral software.

Keywords: String matching algorithm · Smith-Waterman algorithm · JS.Cassandra virus · Polymorphic javaScript virus · Hex and DNA sequences · Automatic signature generation

1 Introduction

Computer malware variants continue to multiply even with the employment of intrusion identification, firewall, and malware identification systems. For a malware identification system, the main problem is to identify a malware variant that is not maintained in its signature library. Modern identification approaches are unable to identify new malware variants until they appear, even when a signature of one variant of that specific malware is accessible [1]. The significance of this library-based weakness will grow as more malware attacks employ polymorphic approaches [2, 3]. Large scale commercial antivirus software systems [2, 4–6] identify malware by employing hash signatures of the most important parts of a known virus inside an antivirus monitor for scanning incoming packets and memory, and also by using heuristic approaches, within a sandbox, to identify malicious activity. Software used in research [2, 7] has concentrated on the automatic generation of viral signatures utilising

© Springer International Publishing Switzerland 2016
M. Chau et al. (Eds.): PAISI 2016, LNCS 9650, pp. 146–165, 2016.
DOI: 10.1007/978-3-319-31863-9_11

multiple strings and regular expressions. These approaches have been reported to be more efficient but are still unable to detect some or all new instances of polymorphic malware, which typically leaves the viral instructions alone but changes the appearance of the virus through changes in the encryption and decryption methods or through reordering of instructions [1]. Other polymorphic techniques include redundant code insertion that does not affect the static parts (i.e. unchanged functions) of the virus. The decrypted viral code is mainly identical for each variant, thereby allowing memory based signature detection of the original virus and its variants: block hashing followed by signature scanning can be effective in identifying memory based traces of polymorphic viruses. Recent studies have concentrated on structural approaches to compare programs through designs of control-flow graph isomorphism [2, 8–10]. However, polymorphic viruses that employ instruction group reorderings can generate non-isomorphic control-flow graphs, restricting the usefulness of graph-based identification techniques [1].

From a theoretical perspective, while virus detection is undecidable [11–13], it is still not known whether it is possible to develop an algorithm that will take an arbitrary program or code excerpt and reliably determine whether it contains specific forms of malware [14]. Because of increasing complexity of obfuscation as well as discovery of new types of modified malware (e.g. spyware, botnets, adware, and ransomware), human experts are still required to implement polymorphic and metamorphic malware detection techniques based on virus behavior and semantics [15–18]. The main problem of such semantically-oriented signature extraction approaches is that an infection must occur first before virus signatures can be identified and extracted manually. Predicting future metamorphic and polymorphic viral forms to prepare AVSs (Anti-Viral Systems) for as yet unseen variants has remained a distant research goal for both semantic and syntactic approaches.

In this study, we will focus on polymorphic malware produced through changes in the encryptor and decryptor. Metamorphic malware, on the other hand, changes its code and therefore the physical appearance of its functions (e.g. code rearrangement through permutation, redundant code expansion and shrinkage) While the static parts of polymorphic viruses such as the virus body can be used to detect polymorphic variants of the same virus despite mutations in the encryptor and decryptor, there may be no static parts to a metamorphic virus. In contrast to polymorphism detection through memory-based signature detection, metamorphic viruses are analysed semantically (e.g. emulated in an isolated sandbox) so that different variants are identified through common functionality and operation. It is difficult to say if polymorphic malware or metamorphic malware is harder to detect. But malware with both polymorphic and metamorphic capabilities will be the hardest malware to detect and protect against [19]. We focus on polymorphic malware detection in this study to identify variants through syntactic analysis alone. Polymorphic malware is comprised of 3 segments: the malware body, the decryptor, and the mutation engine (see Fig. 1). During the infection process, a polymorphic malware M will duplicate itself into a new malware $M' \in L_M$ by selecting an arbitrary encryption key k, encrypting the malware body and then the mutation engine with the key k and ultimately producing a fresh decryptor, implanting k. Since the mutation engine and the malware body are encrypted with an alternate arbitrary key, identifying polymorphic malware usually signifies identifying its

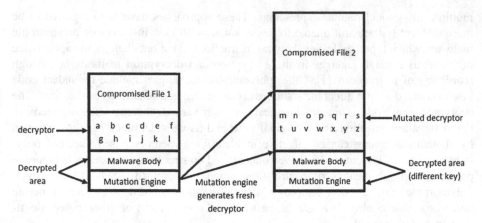

Fig. 1. The process of polymorphic malware infection (Source: NSS, p. 41)

clear-code decryptor. Therefore, for this type of malware, L_M is the decryptor language, rather than the entire malware language [20].

Modern detection techniques are not currently adequate in detecting polymorphic malware for two main reasons. Firstly, the set of all occurrences of a polymorphic malware may define a non-finite state formal language [21] and therefore cannot be identified through regular expressions [20]. Secondly, semantic-based signature extraction approaches can fail to deal with new polymorphic variants because the behavior has changed through possible code rearrangement even if the functionality has not. Current signature extraction is either by manual evaluation or by a learning process that is, as yet, unknown. It has been suggested that learning complicated language classes, like regular or context-free grammars, is not desirable from only positive inputs [22], but it is not known what an ideal negative class of virus should be (e.g. random code, viral code with the payload removed, non-viral programs, etc.). Until now, AVSs have just about kept pace with new variants despite the efforts required to generate signatures manually, probably because polymorphic variants, up till now, demonstrate low levels of complexity. But increasing sophistication of malware writers may soon make this technique unmanageable [20].

The method of forcing a ciphered file to decipher itself is known as generic decryption and it is this technique that is employed by most of the antiviral software products. The main issue with the process of generic decryption is slowness. The process of generic decryption is of no functional use if it takes several hours waiting for a polymorphic malware to decode within the virtual machine. On the other hand, if the process of generic decryption halts too early, there may not be sufficient information to identify a signature [23]. So to overcome this issue, the process of generic decryption uses heuristic techniques - a generic group of instructions that assists in distinguishing non-malware from malware activity with additional information specific to that malware. Even the use of heuristic-based approaches requires progressive analysis and modification. Heuristic-based rules may have to be adjusted to identify 300 types of malware, but may neglect 15 of those malware types when modified to identify 10 new malware types [23]. But the main problem for semantic-based approaches is that the virus or its variant must already be in circulation before it can be identified and detected for removal.

Syntactic approaches to signature extraction based on structural identification of malware are relatively unexplored in comparison to semantic approaches. The historical reason is that the same viral function can be expressed in many different physical code forms and so, the argument goes, only semantic analysis will reveal commonalities among variants of the same virus. This argument may be sound for future, complex malware with sophisticated morphing techniques. Our experience with current and past viruses, however, is that current morphing techniques make simple and unsophisticated changes to generate new variants, and these may be enough to escape detection by semantic-based signatures. Our research hypothesis is that, for current malware, it is possible to identify syntactic structures that help to determine whether a piece of code contains a virus type and its variants. If our research hypothesis does not apply to simple polymorphic variants of a virus, it is unlikely that syntactic structures will be found for more complex polymorphic variants.

What has changed considerably since the historical view that only semantic analysis will reveal viral signatures is the huge growth in our knowledge of string-based search algorithms in bioinformatics during the last 20 years or so as well as growing knowledge of how to categorize viral polymorphism using level (details below). String-based algorithms in bioinformatics do not just search for the presence or absence of characters in certain positions but also manipulate the strings to allow for insertion and deletion of characters to maximize the number of matching characters. Substitution matrices are also used to allow match between non-identical characters if there is a chance of mutation to another character. Such substitution matrices can be built empirically from previous string matches or *a priori* according to predetermined substitution rules. However, previous work [24] has shown that string matching algorithms taken from bioinformatics work best with biologically represented strings (DNA, protein) rather than arbitrary character sets such as hex code. Hence, conversion of viral code to an appropriate biological representation is required before sequence matching, with conversion back to hex code for signature generation.

This huge growth of interest in searching for interesting patterns in biosequences as part of international developments in bioinformatics provides researchers interested in syntactic approaches to virus detection with possibly new techniques and added knowledge when dealing with viruses. The main advantage with a syntactic or structural approach is that new, possible variants can be generated from the extracted syntactic or structural rules of existing viruses, even if those variants do not actually exist, in the same way that new strings can be created even if they never been seen before. The aim of this paper is to explore whether current string searching algorithms of reasonable sophistication and underpinned by increasing knowledge of how to apply the algorithm heuristically for maximum effect, such as the Smith-Waterman algorithm, can lead to genuinely novel syntactic approaches to the automatic generation of viral signatures not just for virus variants already encountered but also for variants not yet encountered. We use the JS.Cassandra virus and its known variants for our experimental purposes. This virus was discovered in 2003 and written by a virus author named 'Second Part to Hell' in Austria. It's a JavaScript virus with a dedicated polymorphic engine (more details below).

2   Systems and Methods

2.1   Technical Safeguards

Downloading all the polymorphic malware and its known variants, as well as hex (hexadecimal) dump extraction and testing were undertaken on a stand-alone system to prevent possible unintended infection of other systems. Network connectivity was used only at the testing stage but was done using 'Oracle VM VirtualBox' (an x86 software package with virtualisation capability) with a pre-installed Linux-based (Ubuntu) operating system image [25]. Due to possible security sensitivity with regard to some of the methods below (*Step-1* and *Step-7*), especially with regard to generating hex dumps from a polymorphic malware executable and its variants, as well as its testing, interested readers are requested to contact the corresponding author, using their academic email addresses, for further information. Our methodology consists of seven steps (see Fig. 2).

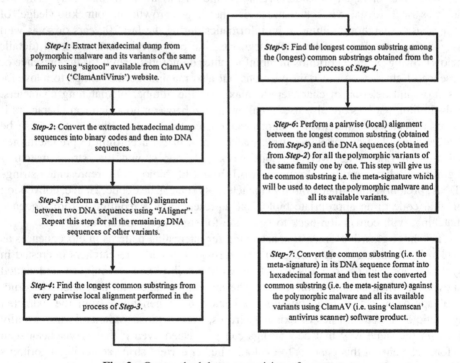

Fig. 2. Our methodology comprising of seven steps

2.2   Hex Dump Extraction

Unlike any other JavaScript virus, JS.Cassandra is comprised of four distinct polymorphic engines: polymorphic engine I, which includes Garbage or Junk codes; polymorphic engine II, which modifies its Body (Body Changing); polymorphic engine III, which modifies its Variables (Variable Changing); and polymorphic engine IV, which modifies its Numbers (Number Changing). The likelihood of the virus to decode using polymorphic engine I is either 1:3 or every 1:4th line (inside its viral code) and engine I has a polymorphism level of 3 (see Table 1). The likelihood of the virus to decode using polymorphic engine II and III are 1:3 and engine II has a polymorphism level of 6, whereas, engine III has a polymorphism level of 2. The likelihood of the virus to decode using polymorphic engine IV is either 1:1 or every 1:10th number (found inside its viral code) and engine IV has a polymorphism level of 2 [26, 27]. The original 'JS.Cassandra' virus with its original source code was downloaded from its original virus author's (Second Part to Hell) website [26]. All the 351 known polymorphic variants of 'JS.Cassandra' virus were also retrieved from its original virus author's website [28].

Table 1. Levels of polymorphism (Source: VX Heavens) [27].

Levels of Polymorphism	Attributes
Level 1 Virus	A virus with a batch of decryptors containing constant code, selecting one while infecting. Such a virus is known as "Oligomorphic" or "Semi-Polymorphic" virus.
Level 2 Virus	A virus with a decryptor containing one or more constant instructions, while the rest of its viral code is modifiable.
Level 3 Virus	A virus with a decryptor containing non-functioning functions or 'junk' such as, STI, NOP, XOR, CLI, etc.
Level 4 Virus	A virus with a decryptor employing interchangeable instructions and modifying its sequence, the process known as instruction mixing. The algorithm of this decryptor remains unmodified.
Level 5 Virus	A virus in this level possess all the attributes mentioned in levels 1-4. The algorithm of the decryptor in such a virus is modifiable and also frequent encoding of the viral code as well as semi-encoding of the decryptor code are feasible.
Level 6 Virus	A virus in this level is known as a permuting virus and its code may be unencrypted.

Step-1: As the 'JS.Cassandra' virus and all its known variants were written in JavaScript programming language, all their source code was readily available. 42 variants out of the 351 known polymorphic malware variants were taken for our experimental testing purpose plus the original 'JS.Cassandra' virus (a total of 43 malicious files). An additional 43 non-malicious/non-payload files were created by removing their primary polymorphic engines. Some of the non-malicious files still had their polymorphic functions in place. A third set of 43 random files were created (i.e. randomly coded) with random sizes in JavaScript file format for testing our methodology against false positives and false negatives. The uniqueness of 43 malicious variants, 43 non-malicious variants and 43 random files was cross-checked by generating its CRC32b hash values (see Table 2). A 32 bit CRC (Cyclic Redundancy Check) i.e. CRC32b hash, provides about 4 billion unique hash values. From Table 2, it can be seen that no two files have the same CRC32b hash values and file size.

Table 2. Generated CRC32b hash values and file sizes in bytes for 43 malicious Files, 43 non-malicious files and 43 random files.

CRC32b Hash Values for 43 Malicious Variants	File Sizes in bytes for 43 Malicious Variants	CRC32b Hash Values for 43 Non-malicious Variants	File Sizes in bytes for 43 Non-malicious Variants	CRC32b Hash Values for 43 Random Generated Files	File Sizes in bytes for 43 Random Generated Files
26489347	7,767	ab657f45	1,823	bfa1e3d9	9,216
848562f1	8,324	3d94f85f	2,662	8956b4ef	4,096
fab48c8c	54,183	634041fe	28	07baad1b	5,120
7c4ea313	9,938	90dd470d	3,697	0ca68128	7,168
bd3b9fdc	8,759	0631e490	2,981	819ec3a5	10,240
9904ef9c	8,392	5273cd32	2,137	048d638c	11,264
511621c7	9,400	32b7909a	3,748	f6425fdb	13,312
a7bc9795	10,059	d1f95eae	2,125	9fda09d6	15,360
a878abc3	10,763	cd486121	2,868	8b2ff426	17,408
ec3797e7	12,282	e2220b79	3,874	e3702e86	20,480
a2e5c540	10,799	d4cffb98	2,965	ae57c29a	19,456
9c8432d2	10,873	91f3e71f	1,688	dbfae5e5	133,120
2b40aa76	8,639	03fe3ba9	1,439	0d4cd9da	52,224
92c87b26	11,334	a6308749	3,021	8a8d3664	35,840
52653b1d	9,507	6eaa1574	1,885	fd11933b	24,576
851c41b7	9,740	e7cb7513	2,872	df99e215	27,648
f006361e	14,900	0b52da18	3,876	fe998b7c	22,528
8ead30b2	9,945	31838f54	2,887	c39cd570	33,792
1ea87480	46,691	505c6653	27	8b461802	45,056
ecd82d26	10,677	db5a61 cc	1,840	79b2dfb3	26,624
59e9feb3	139,909	634041fe	28	d3951707	37,888
70763a3b	14,767	c7b5d591	6,461	68cdd062	103,424
f4289eb6	52,595	26fae117	35,637	12307f18	59,392
c4290e04	29,603	687ff9af	15,230	c0cd6499	52,224
3797337e	25,828	8be6dc4c	9,394	6ec3f157	74,752
4736f8b5	45,659	bb0289a8	22,372	3d5fa9ad	54,272
c2b04d58	45,551	2bcc0d72	21,713	e2f21971	44,032
a181f255	92,807	afcab12f	69,603	384a6e54	49,152
f09e878a	52,166	91a706d4	29,312	210033d2	34,816
1508c8c9	92,418	c08fd2bc	63,980	edecfc3f	13,312
516fb310	45,161	4beeda26	20,518	bad13ac3	32,768
1fb5398a	48,795	fab8097c	22,407	df86781e	43,008
1ce21c33	63,703	1f89ec68	37,841	7f6a41dc	41,984
477f3b8b	73,644	a55be2cf	44,847	4e4e73b6	21,504
e13deddd	101,869	6090639e	69,861	2969861d	23,552
652475dc	108,964	394b9964	76,520	5e26f76b	25,600
6c2eb137	104,588	b9c78d0d	74,971	cafef1e2	28,672
8efd2988	72,866	abc0e54d	41,369	1da5bf28	29,696
0c380868	52,306	7ef94488	25,542	0c200f97	30,720
2fb372eb	88,438	f041406f	55,714	91ea7ac7	31,744
bf934d5b	60,612	2c940492	29,215	2ac33710	36,864
200270d4	79,344	00b715f9	44,847	afc9301b	38,912
70266dfb	103,509	d2d13cbc	70,014	cb672fef	39,936
Total File Size →	1,878,074	**Total File Size** →	935,839	**Total File Size** →	1,482,752

A Java function called 'eval(String.fromCharCode())' has been used in the original JS.Cassandra virus and all its known variants. This function converts the Unicode number into a character. A good example of this function is shown in Fig. 3. The image on the top window has been taken from one of the variant of JS.Cassandra, which purely consists of all numbers and the file size of that variant is 139,909 bytes. This variant when executed converts into a different variant file with a new file size of 9,528 bytes shown in Fig. 3 in the window below, which contains all the normal Java functions. This conversion from one form of variant file into another represents a perfect example of polymorphism.

Fig. 3. Screenshot of JavaScript code comparison resulting from the conversion of one form of polymorphic JS.Cassandra variant file into another

The JavaScript files for all 86 programs were checked using the 'VirusTotal' website to confirm that malicious functionality was preserved in the 43 malicious variants and removed in the 43 non-malicious/non-payload variants. 'VirusTotal' in turn employs 56 well-known antivirus software products and so supplies good assurance that our manual code alterations for non-malicious files were effective. The 43 randomly generated code files were also checked using the 'VirusTotal' website to confirm their uniqueness as it creates a unique SHA256 signature for every file that's uploaded. Hex dumps were then extracted from the 43 malicious and 43 non-malicious variant files using the tools available on the ClamAV ('Clam AntiVirus') website, which uses a program called "sigtool", to generate their hex dumps.

2.3   Hex to DNA Conversion

Step-2: In this step, the extracted hex dump sequences were converted into binary codes and then into DNA sequences. Conversion of hexadecimal into binary code was performed using the following rules: '0' → '0000'; '1' → '0001'; '2' → '0010'; '3' → '0011'; '4' → '0100'; '5' → '0101'; '6' → '0110'; '7' → '0111'; '8' → '1000'; '9' → '1001'; 'a' → '1010'; 'b' → '1011'; 'c' → '1100'; 'd' → '1101'; 'e' → '1110'; and 'f' → '1111'. Subsequent conversion of the binary code into DNA sequences for input to JAligner [29] was performed using the following rules: '00' → 'A'; '11' → 'T'; '10' → 'G'; and '01' → 'C'. An in-house macro tool was designed to perform these conversions. All the 43 extracted hex dumps for malicious files were converted into 43 DNA sequences using this in-house macro tool. All the 43 extracted hex dumps for non-malicious files were also converted into DNA sequences and were kept for use in *Step-3* to *Step-7*. A short example of the conversion of 32-bit binary code into 16 DNA characters is shown below:

10010011111100110001001110011101 (32 − bit binary code)
GCATTGCGAGCTATTC (16 DNA characters)

2.4   Pairwise Local Alignment

The String matching Smith-Waterman algorithm was used here to perform pairwise local alignment. String matching algorithms are applied to identify one or more locations in one string where other strings called patterns are found. Let 'Σ' be a character (an alphabet) i.e. a finite set. Conventionally, both the searched string and pattern are vectors of components of 'Σ'. The 'Σ' may be a regular alphabet i.e. for instance the letters A to Z in the Latin format. Other algorithms may employ binary codes ($\Sigma = \{0,1\}$) and in the area of Bioinformatics, DNA letters ($\Sigma = \{A,T,G,C\}$) [30].

The Smith-Waterman algorithm (SWA) was used to extract the most common substring/pattern from the same family of 43 JS.Cassandra polymorphic malware variants. The SWA conducts local sequence alignment between the two strings to find matching sections between the two strings. The two strings can be a protein or nucleotide sequence. The SWA finds the most matched substrings among the search string and pattern. Rather than observing the complete sequence, the SWA extracts the sections of all feasible length, then compares and enhances the similarity rate. The SWA can look for identical matches or substituted matchers (i.e. a character in the string can be replaced by a different character, including no character (gap), in the pattern, and vice versa). The SWA is a variation of the Needleman-Wunsch algorithm (NWA) [31], both of which are dynamic programming algorithms. Intrinsically, the

SWA is assured to find the optimal local alignment with regard to the scoring method being employed (i.e. the gap scoring and the substitution strategy). There are many substitution matrices available and employed by the SWA such as BLOSUM, IDENTITY (ID) and PAM matrix. However, ID was used in our experiments to perform exact matching. The results of the SWA are called 'alignments', since either or both strings can be changed with gap insertions to produce optimal pattern matches.

Step-3: In this step, a pairwise (local) alignment was performed using the SWA with ID matrix between two generated DNA sequences using a tool called 'JAligner' [29]. 'JAligner' is an open source Java application of the SWA. For instance, if we have 4 variants, say, P1, P2, P3 and P4, a pairwise local alignment will be performed between P1 and P2, followed by P2 and P3 and then P3 and P4, respectively. We applied a similar procedure for all the 43 converted malicious DNA sequences.

Step-4: After the process of local alignment, the longest substrings from every pairwise local alignment were extracted, resulting in 42 longest substrings from the 43 malicious sequences.

Step-5: In this step, the longest common substring among the 42 longest substrings is extracted. This longest common substring represents the longest common DNA substring in the 'family' of 43 JS.Cassandra malware variants. The rest of the longest common DNA substrings were kept for use in *Step-6*.

Step-6: In this step, a pairwise (local) alignment was performed between the longest common DNA substring (obtained from *Step-5*) and the converted 43 malicious DNA sequences (obtained from *Step-2*) one by one. There will be a common substring that will be the same for all the polymorphic malicious variants. This common substring will be the meta-signature that will be used to detect all the known polymorphic variants of that family. If the first longest common substring doesn't give us the desired common substring then repeat the process of *Step-6* using the second longest common substring, if not, then repeat the process of *Step-6* using the third longest common substring, and so on. The common substring (i.e. the meta-signature) of sequence length 80 obtained from this step, is shown below in their DNA representation:

CCATCTCACTAGCGGCCGTGCGCTAGTGCGCGCTAGCGTTCGTCCAATCGGACGACCTAGCAATCG
TTCGCACGCCAGGA

2.5   Meta-Signature Testing

Step-7: In this final step, the common substring in their DNA sequence format was converted into hexadecimal format. The converted hex meta-signature was tested against the JS.Cassandra original virus and all its 42 polymorphic malware variants using ClamAV ('clamscan' antivirus scanner) software. ClamAV is the only known antivirus software product that shows everyone how a malware signature is created and also encourages users to upload their own generated signatures into their database. Basically, ClamAV is an open source antivirus engine that helps in identifying all kinds of malware and it has made the source code of its antivirus tool available publicly [32]. The converted hex meta-signature of sequence length 40 obtained from this step, is shown below:

$$537472696e672e66726f6d43686172436f646528$$

Step-3 to *Step-7* were performed on the 43 non-malicious DNA sequences. The common substring (i.e. the meta-signature) was obtained from *Step-7*, which was exactly the same as the one obtained from the above steps.

3 Results

Table 3 provides the detection ratios of the top 12 well-known antivirus software products (AVSPs) obtained from the 'VirusTotal' website. All the 43 malicious, 43 non-malicious and 43 random files were scanned against the 56 well-known AVSPs but

Table 3. Detection ratios of top well-known antivirus software products for 43 malicious files, 43 non-malicious files and 43 random files obtained from VirusTotal website.

Top AVs		AVG	Avast	Avira	Bitdefender	ClamAV	ESET-NOD32
43 Malicious Files	Detection Rate (Accuracy)	17/43 (39 %)	35/43 (81 %)	12/43 (28 %)	1/43 (2.3 %)	40/43 (93 %)	22/43 (51 %)
	Sensitivity/Recall	39 %	81 %	28 %	2.3 %	93 %	51 %
	Specificity	0.0 %	0.0 %	0.0 %	0.0 %	0.0 %	0.0 %
	Precision	100 %	100 %	100 %	100 %	100 %	100 %
	F1 Score	57 %	90 %	44 %	4.5 %	96 %	68 %
43 Non-Malicious Files	Detection Rate (Accuracy)	0/43 (0.0 %)	0/43 (0.0 %)	0/43 (0.0 %)	0/43 (0.0 %)	0/43 (0.0 %)	0/43 (0.0 %)
	Sensitivity/Recall	0.0 %	0.0 %	0.0 %	0.0 %	0.0 %	0.0 %
	Specificity	100 %	100 %	100 %	100 %	100 %	100 %
	Precision	0.0 %	0.0 %	0.0 %	0.0 %	0.0 %	0.0 %
	F1 Score	0.0 %	0.0 %	0.0 %	0.0 %	0.0 %	0.0 %
43 Random Files	Detection Rate (Accuracy)	0/43 (0.0 %)	0/43 (0.0 %)	1/43 (2.3 %)	0/43 (0.0 %)	0/43 (0.0 %)	0/43 (0.0 %)
	Sensitivity/Recall	0.0 %	0.0 %	0.0 %	0.0 %	0.0 %	0.0 %
	Specificity	100 %	100 %	98 %	100 %	100 %	100 %
	Precision	0.0 %	0.0 %	0.0 %	0.0 %	0.0 %	0.0 %
	F1 Score	0.0 %	0.0 %	0.0 %	0.0 %	0.0 %	0.0 %
Top AVs		Kaspersky	McAfee	Microsoft	Panda	Symantec	Trend Micro
43 Malicious Files	Detection Rate (Accuracy)	1/43 (2.3 %)	22/43 (51 %)	43/43 (100 %)	1/43 (2.3 %)	1/43 (2.3 %)	1/43 (2.3 %)
	Sensitivity/Recall	2.3 %	51 %	100 %	2.3 %	2.3 %	2.3 %
	Specificity	0.0 %	0.0 %	0.0 %	0.0 %	0.0 %	0.0 %
	Precision	100 %	100 %	100 %	100 %	100 %	100 %
	F1 Score	4.5 %	68 %	100 %	4.5 %	4.5 %	4.5 %
43 Non-malicious Files	Detection Rate (Accuracy)	0/43 (0.0 %)	0/43 (0.0 %)	0/43 (0.0 %)	0/43 (0.0 %)	0/43 (0.0 %)	0/43 (0.0 %)
	Sensitivity/Recall	0.0 %	0.0 %	0.0 %	0.0 %	0.0 %	0.0 %
	Specificity	100 %	100 %	100 %	100 %	100 %	100 %
	Precision	0.0 %	0.0 %	0.0 %	0.0 %	0.0 %	0.0 %
	F1 Score	0.0 %	0.0 %	0.0 %	0.0 %	0.0 %	0.0 %
43 Random Files	Detection Rate (Accuracy)	0/43 (0.0 %)	0/43 (0.0 %)	0/43 (0.0 %)	0/43 (0.0 %)	0/43 (0.0 %)	0/43 (0.0 %)
	Sensitivity/Recall	0.0 %	0.0 %	0.0 %	0.0 %	0.0 %	0.0 %
	Specificity	100 %	100 %	100 %	100 %	100 %	100 %
	Precision	0.0 %	0.0 %	0.0 %	0.0 %	0.0 %	0.0 %
	F1 Score	0.0 %	0.0 %	0.0 %	0.0 %	0.0 %	0.0 %

results for only the top ones are shown in Table 3. According to an article from 'TopTenReviews' website about '2015 Best Antivirus Software Review, 'Bitdefender' is the top rated antivirus with a performance rating' (PR) of 9.73/10 and ranked no.1, followed by, 'Kaspersky' with a PR of 10/10 and ranked no.2, followed by, 'McAfee' with a PR of 9.73/10 and ranked no.3. Other AVSPs listed in Table 3 were, 'Symantec Norton' with a PR of 8.9/10 and ranked no.4, followed by, 'Trend Micro' with a PR of 9.43/10 and ranked no.5. 'Avira' was ranked no.6 (PR of 9.73), 'ESET NOD32' was ranked no.11 (PR of 8.03), 'Panda' was ranked no.12 (PR of 8.05), 'Avast' was ranked no. 14 (PR of 8.03), and 'AVG' was ranked no. 15 (PR of 8.32), were also in the list of top 12 well-known antivirus software products (see Table 3) [33].

From Table 3, it can be seen that only 'Microsoft' antivirus can successfully detect all the 43 malicious JS.Cassandra variants. Surprisingly, 'Microsoft' antivirus was not listed in the 2015 best list of 'TopTenReviews', which could detect 100 % of the 43 highly polymorphic malware variants. 'Bitdefender' being the top ranked antivirus could only detect 1 out of the 43 malicious malware variants with a lowest accuracy of 2.3 %. Even the second and third top rated antivirus software products (i.e. 'Kaspersky' and 'McAfee'), could only detect 1 out of the 43 (accuracy of 2.3 %) and 22 out of the 43 (accuracy of 51 %) malware variants, respectively. 'ClamAV' had the second highest detection ratio i.e. 40 out of 43 (accuracy of 93 %), but again was not found on the 2015 best antivirus list. Third highest was the 'Avast' antivirus, with a detection ratio of 35 out of 43 (accuracy of 81 %) and is ranked no. 14 on the best 2015 list. 'Avira' antivirus, which is ranked no. 6, and the only antivirus that detected a random generated file giving us a false positive of 1 out of 43 (i.e. specificity of 98 %).

The meta-signature obtained from *Step-7* was tested against the 43 malicious, 43 non-malicious and 43 random files. A database file was created in .ndb file format (ClamAV Extended Signature File) for testing the meta-signature using the 'clamscan' antivirus scanner. The database format for 'clamscan' goes specifically in the following format i.e., 'MalwareName: TargetType: Offset: HexSignature' [34]. Figures 4, 5 and 6

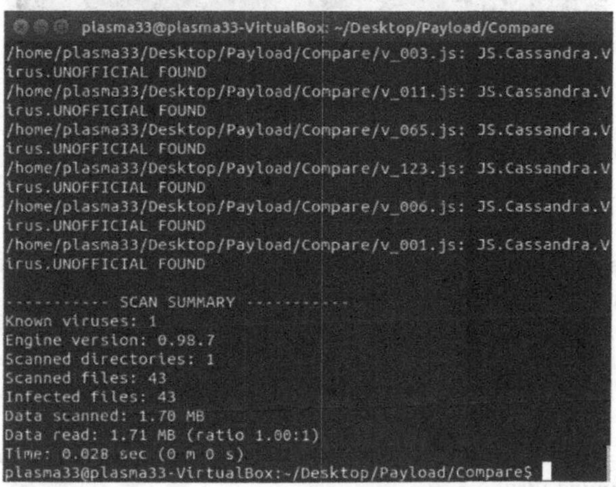

Fig. 4. Screenshot of the scan result obtained from 'clamscan' antivirus scanner for 43 malicious files using the meta-signature

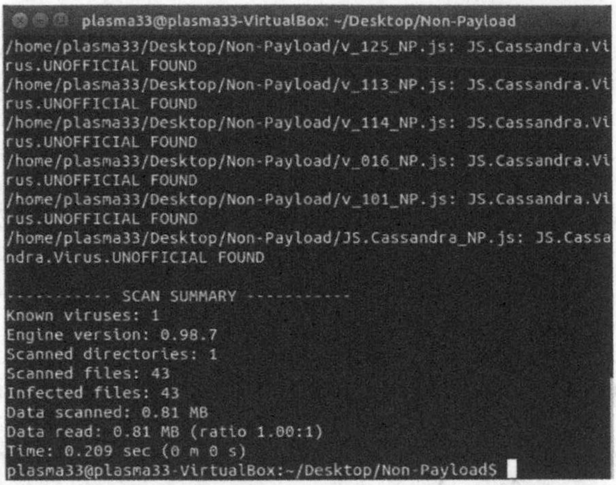

Fig. 5. Screenshot of the scan result obtained from 'clamscan' antivirus scanner for 43 non-malicious files using the meta-signature

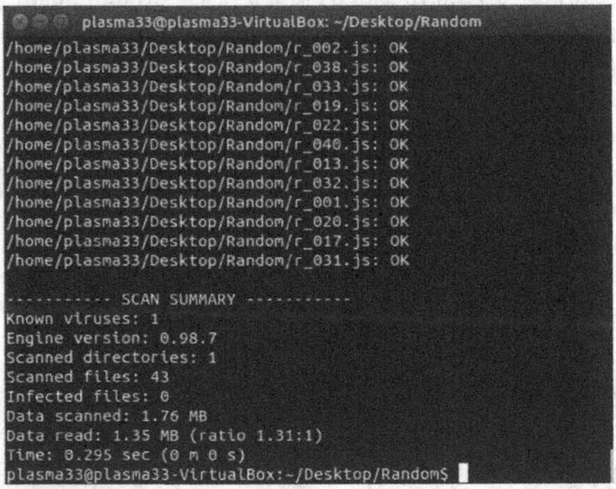

Fig. 6. Screenshot of the scan result obtained from 'clamscan' antivirus scanner for 43 random generated files using the meta-signature

are the screenshots of the scan results obtained from 'clamscan' antivirus scanner for 43 malicious, 43 non-malicious and 43 random files using the meta-signature obtained from the 7 step approach. The content inside the .ndb database file is shown below:

JS.Cassandra.Virus: 0 : * : 537472696e672e66726f6d43686172436f646528

Where, 'JS.Cassandra.Virus' is the common polymorphic family malware filename for the JS.Cassandra virus.

'0' is the 'TargetType' (i.e. the type of target file in our case it's a JavaScript file) and several options are available within which are, '0' is for any file, '1' is for portable executable file, '2' is OLE2 component i.e. a vb script file, '3' is for normalised HTML, '4' is for mail type files, and '5' is for graphics files. Option '0' was chosen in this case [34].

'*' is the 'Offset' type to tell the scanner about where the signature applies inside the file (similar to {n} wildcard) & 3 options are available within, which are, '*' is for anywhere inside the file, 'n' is for n bytes from beginning of file, & 'EOF – n' is for End Of File minus the n bytes. Option '*' was chosen in this case [34].

'537472696e672e66726f6d43686172436f646528' is the meta-signature (in hexadecimal format) obtained from the 7 step approach.

Figures 4, 5 and 6 shows that 43 out of the 43 malicious files, 43 out of the 43 non-malicious files and 0 out of the 43 randomly generated JavaScript files were successfully detected as infected by the 'clamscan' antivirus scanner using the meta-signature in 0.028 s, 0.209 s and 0.295 s, respectively. The accuracy of 'clamscan' antivirus scanner using the meta-signature for the original JS.Cassandra polymorphic file and its 43 known malicious as well as non-malicious variant files, was 100 %. In Fig. 4, 5 and 6, under 'SCAN SUMMARY', 'Known Viruses' means the number of signatures found inside the database file (.ndb), which in this case, was only 1 (i.e. the meta-signature).

This meta-signature not only detected 43 malicious files successfully, but also detected 43 non-malicious files. Non-malicious files still had some polymorphic functions intact inside. All the 43 non-malicious files were still executable but a few gave JavaScript run-time and compilation errors. These executable non-malicious files might cause some serious potential threats, as the polymorphic functions inside these files might still make them polymorphic, in some cases. Table 3 shows that none of the 56 AVSPs detected these executable non-malicious files as malicious. One instance of such a file will be explained here.

A variant of JS.Cassandra was chosen and around 4 polymorphic functions were removed from it. Still more than 10 functions were intact inside that variant file. The modified variant file was executed and 12 new unique polymorphic variants were generated from it for experimental purpose of distinguishing between malicious and non-malicious files. An infinite number of new unique polymorphic variants could be generated from the same file. Table 4 provides the CRC32b hash values and file sizes in bytes for the original variant file (V_O), modified variant file (V_M) and 12 new variants (V_{M1}-V_{M12}). From Table 4, it can be seen that no two files have the same CRC32b hash values and file size.

Table 5 provides the detection capabilities of top 12 well-known AVSPs obtained from the 'VirusTotal' website. All 12 new variant files (V_{M1}-V_{M12}), original variant file (V_O) and modified original file (V_M) were scanned against the 56 well-known AVSPs but results for only the top ones are shown in Table 5. From Table 5, it could be said that only 'Microsoft' antivirus could successfully detect all the 14 malicious files. 'Bitdefender' and 'Kaspersky' being the top 2 rated AVSPs couldn't detect any of those 14 malicious files. 'McAfee', which is ranked third, could only detect 1 out of the 14 malicious files. It could only identify the original variant file V_O. Original variant

Table 4. Generated CRC32b hash values and file sizes in bytes for original variant file (V_O), modified variant file (V_M) and 12 variants (V_{M1}-V_{M12}).

Malicious Files	CRC32b Hash Values	File Sizes in bytes
Original Variant File V_O	848562f1	8,324
Modified Variant File V_M	fc79adfe	5,695
Variant 1 V_{M1}	557562ad	6,289
Variant 2 V_{M2}	150e0d7a	6,855
Variant 3 V_{M3}	5645b651	7,457
Variant 4 V_{M4}	42f590e7	7,465
Variant 5 V_{M5}	fcb28864	8,049
Variant 6 V_{M6}	dd679959	8,055
Variant 7 V_{M7}	204f3304	9,019
Variant 8 V_{M8}	3e2ef86f	9,649
Variant 9 V_{M9}	3d987ac4	10,713
Variant 10 V_{M10}	0ecba96f	11,255
Variant 11V_{M11}	bbe2b767	12,825
Variant 12 V_{M12}	55a47fbf	14,031
–	**Total File Size →**	125,681

file V_O before the modification could be detected by 5 antivirus software products but after the modification, V_M could only be detected by the 'Microsoft' antivirus.

Now, the meta-signature (obtained from *Step-7*) was tested to detect those 14 new malicious files. Figure 7 shows that all the 14 files were successfully detected as infected by the 'clamscan' antivirus scanner using the meta-signature in 0.008 s.

The original JS.Cassandra polymorphic virus file and its 351 known variant files were scanned for malicious activity using 'Microsoft' antivirus, 'ESET' antivirus (universally known antivirus software product) both installed on a Windows based system and 'clamscan' antivirus scanner installed on a Linux based system using their own ClamAV database and using our database containing the meta-signature. The databases of all the antivirus software products were up-to-date with the latest updates. Figure 8 shows that all the 351 variant files as well as the original JS.Cassandra file were successfully detected as infected by the 'clamscan' antivirus scanner software product using the meta-signature in 0.995 s. Table 6 shows that only 'Microsoft' antivirus and detection using the meta-signature could successfully identify all the malicious files with a detection ratio of 352 out of 352 (accuracy of 100 %).

Detection of two other highly polymorphic malware and its known variants were also tested using our meta-signature and the results were 100 % successful (see Table 6). Although, some AVSPs could successfully identify all the malicious files but were not consistent in identifying all the highly polymorphic malware and its known variants shown in Table 6. 'Microsoft' antivirus could only identify 80 out of the 1106 'Win32.Kitti' malicious files with an accuracy of 7 % but could successfully identify all the malicious files for the other 2 polymorphic malware.

Table 5. Detection capabilities of top well-known antivirus software products for original variant file (V_O), modified variant file (V_M) and 12 new variants (V_{M1}-V_{M12}).

Top AVs	AVG	Avast	Avira	Bitdefender	ClamAV	ESET-NOD32
Original Variant File (V_O)	Yes	No	No	No	Yes	Yes
Modified Variant File (V_M)	No	No	No	No	No	No
12 Variant Files (V_{M1}-V_{M12})	No	No	No	No	No	No
Top AVs	**Kaspersky**	**McAfee**	**Microsoft**	**Panda**	**Symantec**	**Trend Micro**
Original Variant File (V_O)	No	Yes	Yes	No	No	No
Modified Variant File (V_M)	No	No	Yes	No	No	No
12 Variant Files (V_{M1}-V_{M12})	No	No	Yes	No	No	No

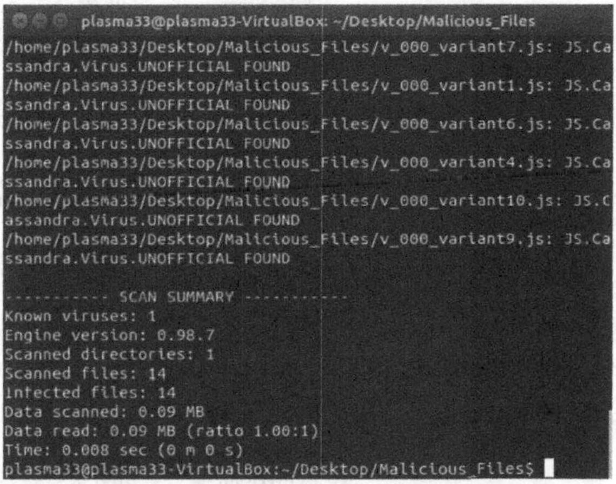

Fig. 7. Screenshot of the scan result obtained from 'clamscan' antivirus scanner for 12 variant files (V_{M1}-V_{M12}), original variant file (V_O) and modified variant file (V_M)

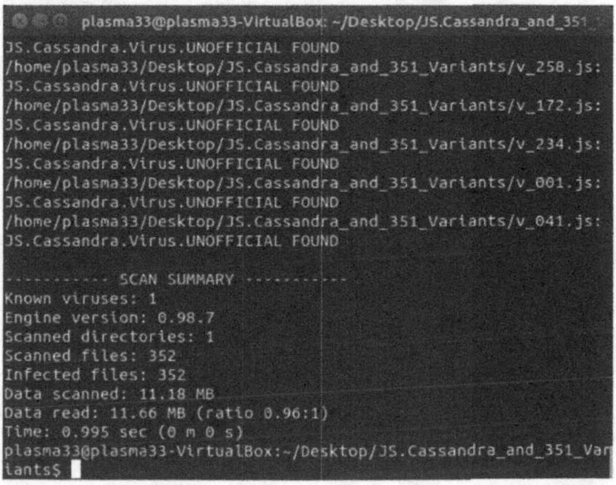

Fig. 8. Screenshot of the scan result obtained from 'clamscan' antivirus scanner for 352 malicious files of JS.Cassandra

Table 6. Detection ratio and time duration for the detection of 3 highly polymorphic malware and its known variants using ESET, clamscan (ClamAV Scanner), microsoft and our meta-signature.

Polymorphic Malware	Antivirus Software Products	Detection Ratio		Time Duration for the Detection
JS.Cassandra Virus and its 351 Variants	**Microsoft**	**Detection Rate** (Accuracy)	352/352 (100 %)	Not Available
		Sensitivity/Recall	100 %	
		Specificity	0.0 %	
		Precision	100 %	
		F1 Score	100 %	
	ESET	**Detection Rate** (Accuracy)	296/352 (84 %)	4 s
		Sensitivity/Recall	84 %	
		Specificity	0.0 %	
		Precision	100 %	
		F1 Score	91 %	
	ClamScan (ClamAV AntiVirus)	**Detection Rate** (Accuracy)	340/352 (97 %)	30.613 s
		Sensitivity/Recall	97 %	
		Specificity	0.0 %	
		Precision	100 %	
		F1 Score	98 %	
	Our Meta-Signature	**Detection Rate** (Accuracy)	**352/352 (100 %)**	**0.995 s**
		Sensitivity/Recall	**100 %**	
		Specificity	0.0 %	
		Precision	**100 %**	
		F1 Score	**100 %**	
Win32.Kitti Virus and its 1105 Variants	**Microsoft**	**Detection Rate** (Accuracy)	80/1106 (7 %)	Not Available
		Sensitivity/Recall	7 %	
		Specificity	0.0 %	
		Precision	100 %	
		F1 Score	13 %	
	ESET	**Detection Rate** (Accuracy)	1106/1106 (100 %)	43 min and 23 s
		Sensitivity/Recall	100 %	
		Specificity	0.0 %	
		Precision	100 %	
		F1 Score	100 %	
	ClamScan (ClamAV AntiVirus)	**Detection Rate** (Accuracy)	1/1106 (0.09 %)	39.378 s
		Sensitivity/Recall	0.09 %	
		Specificity	0.0 %	
		Precision	100 %	
		F1 Score	0.18 %	
	Our Meta-Signature	**Detection Rate** (Accuracy)	**1106/1106 (100 %)**	**14.003 s**
		Sensitivity/Recall	**100 %**	
		Specificity	0.0 %	
		Precision	**100 %**	
		F1 Score	**100 %**	

(Continued)

Table 6. *(Continued)*

Polymorphic Malware	Antivirus Software Products	Detection Ratio		Time Duration for the Detection
2 x Win32.Cholera Viruses and its 198 Variants	**Microsoft**	**Detection Rate (Accuracy)**	200/200 (100 %)	Not Available
		Sensitivity/Recall	100 %	
		Specificity	0.0 %	
		Precision	100 %	
		F1 Score	100 %	
	ESET	**Detection Rate (Accuracy)**	200/200 (100 %)	23 s
		Sensitivity/Recall	100 %	
		Specificity	0.0 %	
		Precision	100 %	
		F1 Score	100 %	
	ClamScan (ClamAV AntiVirus)	**Detection Rate (Accuracy)**	67/200 (33 %)	34.009 s
		Sensitivity/Recall	33 %	
		Specificity	0.0 %	
		Precision	100 %	
		F1 Score	50 %	
	Our Meta-Signature	**Detection Rate (Accuracy)**	**200/200 (100 %)**	1.008 s
		Sensitivity/Recall	**100 %**	
		Specificity	0.0 %	
		Precision	**100 %**	
		F1 Score	**100 %**	

4 Conclusion

The meta-signature obtained from the 7 step approach for JS.Cassandra and its variants was decoded into `'String.fromCharCode('` which is a JavaScript function. This function will be a common function inside the source code for the original JS.Cassandra polymorphic virus and its known variants. As these viruses don't readily come with an open source code (in case of JS.Cassandra polymorphic virus and its known variants, the source code was openly available but usually, JavaScript based viruses are either encrypted or password protected), it's a very complicated process for malware researchers to reverse engineer the malicious file by retrieving their source code based on its malicious activity. The antivirus scanners utilises the approach of 'Variable Scan Sequences'. While looking for a potential malware, the antivirus scanners searches for string bytes in different positions but in a constant string. This is all carried out within an emulator or in a virtual machine. When one deploys any antivirus program and starts to scan files for malware, what actually eventuates is each individual file added in this scanner is in fact executing within an emulator generated by RAM. Files added in this emulator performs as if executing on an actual machine. The scanner checks and manages the file as it runs within the virtual machine. As noted earlier, this method of forcing an encoded file to decode itself is known as generic decryption but the crucial issue with this method is that it's too time-consuming [23].

The novel syntactic approach (with the aid of string matching SWA) to the automatic generation of viral meta-signatures detected all the known virus variants of JS. Cassandra polymorphic virus (see Tables 3 and 4). ESET cannot successfully detect all the known variants of JS.Cassandra Virus family. Moreover, ClamAV could hardly detect the variants of Win32.Kitti virus family but detected around 33 % of variants for the Win32.Cholera virus family. The selected 3 highly polymorphic virus families were at least 5-11 years old. But as our experiments show, modern antivirus software products cannot successfully detect all the known variants of the polymorphic malware family mentioned here. Serious concerns exist as to whether current AVS technology will detect new (future) variants of polymorphic malware. Our research shows that there is a need for a good detecting software system that can effectively and efficiently detect potentially old, current and future malware variants. The ultimate goal would be to detect all new (future) polymorphic variants using a syntactic approach to identify variants both within a virus family as well as across virus families.

References

1. Thompson, G.R., Flynn, L.A.: Polymorphic malware detection and identification via context-free grammar homomorphism. Bell Labs Tech. J. Inf. Technol./Netw. Secur. **12**(3), 139–147 (2007)
2. Kruegel, C., Kirda, E., Mutz, D., Robertson, W., Vigna, G.: Polymorphic worm detection using structural information of executables. In: Proceedings of 8th International Symposium on Recent Advances in Intrusion Detection, pp. 207–226. IEEE (2005)
3. VX Heaven. (2015) VX Heavens Library, 3 May 2015. http://vxheaven.org/
4. Kaspersky Anti-virus 6.0. Kaspersky Lab (2005). http://www.kaspersky.com/about
5. Advanced Virus Detection Scan Engine and DATs: Comprehensive Scanning Technology for Today's Threats and Tomorrow's. Network Associates Technology (2002). http://repo.hackerzvoice.net/
6. Understanding Heuristics Symantec's Bloodhound Technology. Symantec (1997). https://www.symantec.com/
7. Newsome, J., Karp, B., Song, D.: Polygraph: automatically generating signatures for polymorphic worms. In: Proceedings of IEEE Symposium on Security and Privacy, pp. 226–241. IEEE (2005)
8. Dullien, T., Rolles, R.: Graph-based comparison of executable objects. In: Proceedings of Symposium sur la Securite des Technologies de l'Information et des Communications, SSTIC (2005)
9. Flake, H.: Structural comparison of executable objects. In: Proceedings of IEEE Conference on Detection of Intrusions and Malware and Vulnerability Assessment, pp. 161–173. IEEE (2004)
10. Sabin, T.: Comparing Binaries with Graph Isomorphisms. SecuriTeam (2004). http://www.securiteam.com/
11. Cohen, F.B.: Computer viruses: theory and experiments. Comput. Secur. **6**(1), 22–35 (1987)
12. Cohen, F.B.: Computational aspects of computer viruses. Comput. Secur. **8**(4), 325–344 (1989)
13. Adleman, L.M.: An abstract theory of computer viruses. In: Goldwasser, S. (ed.) CRYPTO 1988. LNCS, vol. 403, pp. 354–374. Springer, Heidelberg (1990)

14. Zuo, Z., Zhou, M.: Some further theoretical results about computer viruses. Comput. J. **47** (6), 627–633 (2004)
15. Robiah, Y., Rahayu, S., Zaki, M., Shahrin, S., Faizal, M.A., Marliza, R.: A new generic taxonomy on hybrid malware detection technique. Int. J. Comput. Sci. Inf. Secur. **5**(1), 56–60 (2009)
16. Fukushima, Y., Sakai, A., Hori, Y., Sakurai, K.: A behaviour based malware detection scheme for avoiding false positive. In: Proceedings of 6th IEEE Workshop on Secure Network Protocols (NPSec), pp. 79–84. IEEE (2010)
17. Elhadi, A.A.E., Maarof, M.A., Osman, A.H.: Malware detection based on hybrid signature behaviour application programming interface call graph. Am. J. Appl. Sci. **9**(3), 283–288 (2012)
18. Idika, N., Mathur, A.P.: A survey of malware detection techniques. Technical report 286, Department of Computer Science, Purdue University, USA, 7 July 2014 (2007). http://www.serc.net/
19. Skoudis, E., Zeltser, L.: Malware: Fighting Malicious Code. Prentice Hall Professional, Upper Saddle River (2004)
20. Chaumette, S., Ly, O., Tabary, R.: Automated extraction of polymorphic virus signatures using abstract interpretation. In: Proceedings of the Network and System Security, pp. 41–48. NSS (2011)
21. Filiol, E.: Metamorphism, formal grammars and undecidable code mutation. Int. J. Comput. Sci. **2**, 70–75 (2007)
22. Gold, E.: Language identification in the limit. Inf. Control **5**, 447–474 (1967)
23. The Art of Stealthy Viruses (2006) Hackerz Voice, 27 April 2015. http://repo.hackerzvoice.net/depot_madchat/vxdevl/library/The%20Art%20of%20Stealthy%20Viruses.txt
24. Naidu, V., Narayanan, A.: Further experiments in biocomputational structural analysis of malware. In: 10th International Conference on Natural Computation. ICNC, pp. 605–610 (2014)
25. Oracle VM VirtualBox (2015) VirtualBox, 10 March 2014. https://www.virtualbox.org/
26. JS.Cassandra by Second Part To Hell (2015) rRlF#4 (Redemption), 9 March 2015. http://spth.virii.lu/rrlf4/rRlf.13.html
27. Tutorials– Win32 Polymorphism (2014) VX Heavens, 10 March 2015. http://vxheaven.org/lib/static/vdat/tuwin32p.htm
28. Viruses: Second Part To Hell's Artworks – VIRUSES (2004), 10 March 2015. http://spth.virii.lu/Cassandra-testset.rar
29. JAligner (2010) JAligner: Java Implementation of the Smith-Waterman algorithm for biological sequence alignment – SourceForge. 1 May 2015. http://jaligner.sourceforge.net/
30. Charras, C., Lecroq, T.: Exact String Matching Algorithms. Univ. de Rouen (1997), 30 April 2015. http://www-igm.univ-mlv.fr/~lecroq/string/index.html
31. Smith, T.F., Waterman, M.S.: Identification of common molecular subsequences. J. Mol. Biol. **147**, 195–197 (1981)
32. ClamAV Source Code Download (2014) ClamAV, 10 March 2014. http://www.clamav.net/download.html
33. Top 10 Best Antivirus Software for 2015 – Top Ten Reviews (2015) TopTenReviews, 10 September 2015. http://anti-virus-software-review.toptenreviews.com/v2/
34. Create Your Own Anti-Virus Signatures with ClamAV (2008) Adam Sweet's Blog, 26 February 2015. http://blog.adamsweet.org/ and http://www.clamav.net/

Small State Acquisition of Offensive Cyberwarfare Capabilities: Towards Building an Analytical Framework

Daniel Hughes[✉] and Andrew Colarik

Massey University, Albany, New Zealand
Daniel.Hughes.1@uni.massey.ac.nz, A.M.Colarik@massey.ac.nz

Abstract. This paper examines the factors that motivate small states to acquire Offensive Cyberwarfare Capabilities (OCWC) and identifies the circumstances under which acquiring such capabilities is advantageous to a small state. First, the paper will offer a comprehensive analysis of the characteristics and limitations of OCWC, arguing that military conflicts are unlikely to be won solely by cyber weapons. Second, it analyses potential and likely uses of OCWC by small states and how these may advance political objectives, as explained by conceptual security models. Finally, the paper presents the first iteration of an analytic framework designed to provide a customized estimate of the desirability of OCWC acquisition for individual small states. The model is demonstrated by a case study on a member of the Five Eyes intelligence network and quintessential small state: New Zealand.

Keywords: Cyber · Warfare · Small states · Capabilities · Weapons · Acquisition · Framework · New Zealand

1 Introduction

This paper examines the factors that motivate small states to acquire Offensive Cyberwarfare Capabilities (OCWC) and identifies the circumstances under which acquisition would be beneficial. Literature to date has tended to focus on cybersecurity rather than cyberwarfare [1]. Accordingly, there is no analytical framework through which to consider whether small states should invest in OCWC. Questions regarding cyberweapon acquisition by small states will become increasingly important as they face difficult security investment choices due to the escalating costs of military platforms, uncertainties about the capabilities of cyberweapons, and the perception that OCWC are a low cost alternative to traditional military capabilities.

This paper first offers a comprehensive analysis of OCWC, based on a definition of cyberwarfare that, building on existing literature, underscores how cyberwarfare is the extension of policy via actions carried out through the increasingly militarized domain of cyberspace to create kinetic effects comparable to traditional military capabilities. This analysis is complemented by an examination of the balance of power between offensive and defensive cyberwarfare, the limitations of OCWC, and the concept of cyberpower, demonstrating that OCWC can neither win military conflicts unaided nor alter fundamental principles of warfare. Second, it analyses the likely uses of OCWC

© Springer International Publishing Switzerland 2016
M. Chau et al. (Eds.): PAISI 2016, LNCS 9650, pp. 166–179, 2016.
DOI: 10.1007/978-3-319-31863-9_12

by small states and the benefits and risks such use may generate. This analysis begins with a clarifying definition of the term 'small state', then continues with an examination of potential uses of OCWC: warfighting, coercion, deterrence, and defense diplomacy. This analysis is then refined by an examination of how these uses of OCWC can advance small state political objectives, as explained via multiple conceptual small state security models.

Lastly, this paper presents the first iteration of an analytic framework designed to recommend whether a particular small state should acquire OCWC. The framework begins by examining a small state's key quantitative and qualitative characteristics, along with its security and defense policies, military capabilities, and technical, financial and intelligence resources. This information is enriched by a consideration of the small state's 'cyber-dependence'– the degree to which its economy, military, and government rely on cyberspace, then a behavioral analysis of the state and its potential use of OCWC under conceptual small state security models. The sum of this analysis is evaluated against each category of potential OCWC use, resulting in predictive information regarding the utility of OCWC to the small state in question and a recommendation on the overall desirability of OCWC acquisition.

2   The Emergence and Characteristics of Offensive Cyberwarfare Capabilities

Cyberwarfare has become possible due to the advent of cyberspace, which despite its importance, does not have a commonly accepted definition. Building on academic [2], military [3] and policy [4] based definitions, this paper defines cyberspace as *a notional environment that consists of virtual and physical components. Its primary purpose is the transfer, storage and manipulation of information. It is a human-made domain and its existence relies on human-made objects and the energies of the electromagnetic spectrum.*

Cyberspace is the fifth domain (after land, sea, air, and space) to be militarized. These domains are interdependent; activities in one domain can create effects in and through one another [4]. Evidence of the escalating militarization of cyberspace can be seen in strategic documents, increasing investment in OCWC, and how cyber-attacks on US assets can now be considered to be of sufficient severity to warrant a traditional military response [5]. Despite this, cyberwarfare remains a contested term [6, 7]. In this paper, based on a synthesis of existing literature, cyberwarfare is defined as *an extension of policy via the military exploitation of cyberspace to create kinetic effects that approximate the effects of conventional weaponry. These effects either constitute a serious threat to a nation's security, or are conducted in response to a perceived threat against a nation's security* [6, 8, 9]. Accordingly, this paper defines Offensive Cyberwarfare Capabilities (OCWC) as *cyberweapons possessed by military or para-military organizations who have the will and expertise needed to use them to create military-grade kinetic effects.* The authors present these definitions in order to conceptualize this paper's use of OCWC.

Several commentators [4, 10] believe that cyberwarfare strongly favors OCWC over defensive cyber capabilities. Cyberspace, after all, is a target rich environment based on network structures that privilege ease of use over security. Attacks can be launched almost instantaneously; range and location are not limiting factors. There is rapid, growth in the number of networks and assets requiring protection and numerous vulnerabilities within critical infrastructure [11]. There are also considerable technical and legal difficulties that make accurate attribution of, and accurate and proportionate retaliation to, cyber-attacks a fraught process [12]. Finally there is the low cost of creating OCWC. Computer code is inexpensive to produce, and any cyberweapon released into the internet can be adapted to form the basis of new weapons [13].

The established modality of offensive cyber capabilities, however, are in question; other commentators [13, 14] are less certain of the dominance of OCWC. For example, cyber-dependence, the degree to which an attacker is dependent on cyberspace for functioning infrastructure, is crucial: increased cyber-dependence vis-a-vis an opponent will reduce the effectiveness of OCWC and increase vulnerability to retaliation. Uncertainty also rules in cyberwarfare, as shown by the 'dual use' [9] nature of cyberweapons - they can be captured, adapted and turned against their creators. Furthermore cyberattack vectors are rarely direct, and actions taken by states accidentally targeted are a significant risk. Equally important is the concept of 'escalation dominance' [15]. As shown by yet untested US policy, retaliation to a cyberattack need not be limited to cyberspace, but could instead be delivered by more traditional military means. Moreover, while the speed of a cyberattack may be near instantaneous, the preparation for large-scale, sophisticated cyberattacks is considerable. The Stuxnet attack, for example, required expansive espionage, industrial testing, sophisticated code, and clandestine delivery. Its creation required the resources of a technologically sophisticated nation; it was not built overnight [13].

The above illustrates an argument made by Rid & McBurney [16]: "Maximizing the destructive potential of a cyberweapon is likely to come with a double effect: it will significantly *increase* the resources, intelligence and time required to build and deploy such weapons – and more destructive potential will significantly decrease the number of targets, the risk of collateral damage and the coercive ability of cyberweapons." While this statement may seem in opposition to the low cost of creating cyberweapons; the costs it emphasizes are related to targeting and deploying weapons, not *creating* them. Advanced weapons must be targeted *before they are developed*. States must be certain about the objective and target of cyberweapons – they cannot be easily retargeted to meet unforeseen threats.

While OCWC have considerable destructive potential, they do have limitations. Ultimately they are pieces of computer code that rely on exploiting vulnerabilities caused by reliance on cyberspace [17]. They can attack vulnerable platforms and infrastructure by manipulating computer controlled safety systems, or act as a force multiplier to traditional military assets. These effects, however, are always secondary – cyberweapons cannot directly kill, injure, manipulate or destroy without a device to act through, nor can they occupy and control territory. As such, conflicts are unlikely to be won in the cyberspace alone.

Regardless of their limitations, a significant amount has been invested into the development of OCWC [4], and a significant number of states 'include cyberwarfare in their military planning and organization' [18], though reliable data on who possesses which cyberweapons and the capability of these weapons is highly classified [12]. This secrecy creates a broad spectrum of judgement concerning the threat posed by cyberweapons, which varies from conservative [19, 20], to moderate [13], to catastrophic [21]. These perspectives vary according to two factors: how much damage will accompany the compromise of cyber-dependent platforms and the extent to which major disruptions to state capabilities erode political will and can be exploited by traditional military force.

The growth of OCWC has seen some analysts [22, 23] explore the concept of 'cyberpower'. In the context of warfare, 'cyberpower' is only a new source of power in that it arises from a new military domain; it does not change the nature of power – the capacity to modify the behavior of others while preventing others from affecting one's own behavior [24]. Thus while it is important to identify what is new about cyberwarfare, it should be emphasized that cyberwarfare will not replace other domains of warfare, nor will it alter core principles of warfare, which remain subservient to political objectives.

3   Offensive Cyberwarfare Capabilities and Small States

Before analysis can begin on small state acquisition of OCWC, it is necessary to identify what the term 'small state' refers to. There has been no widely accepted definition of what constitutes a small state; disagreements have hindered consistent use of the term in literature [25, 26]. This has led to the rejection of the term [27] due to the relational and contextual nature upon which any classification of states into categories such as 'small', 'medium', or 'large' would rest. Modifying the work by Rickli [25], the characteristics that identify small states fall into three categories of measurement: *quantitative*, *qualitative-behavioral*, and *qualitative-self-identification*. *Quantitative* measures refer to quantifiable measures such as land area, population and Gross Domestic Product (GDP). *Qualitative-behavioral* measures concern the behavior of a state within the international system and *qualitative-self-identification* measures focus on how a state perceives its own identity.

For the purposes of this paper, the categorical tensions between quantitative, qualitative-behavioral, and qualitative-self-identification measures do not need to be resolved. Nor is a single, essential definition of 'small state' required. Instead, drawing on Wittgenstein's concept of 'family resemblance' [28], the concept of a 'small state' can be defined by possession of a sufficient number of overlapping characteristics – some quantitative, some qualitative-behavioral, and some qualitative-self-identification. No one category of measurement is essential to the definition of small state. Thus in order to analyze the acquisition of offensive-cyber-warfare-capabilities, the determination of state size can be made through a contextual and individualized examination of each state in question. For example the same quantitative measures of a state – population, geographic area, may indicate it should be considered as 'small'. However, an advanced economy, well developed military power, a history of exerting international influence and self-identification as a regional power may mean that the state in question should be considered 'medium', not 'small'. This method of classification is representative of the information presented regarding the small state example that is presented in Sect. 4; it is a foundational aspect of the analytical framework offered in this paper.

To understand the benefits derived from small state acquisition of OCWC, it is necessary to understand how OCWC can be used. A non-exhaustive list of potential uses include *warfighting, coercion, deterrence,* and *defense diplomacy*. As OCWC are limited to secondary effects they have limited uses in warfighting. Their most prominent use is the disruption and manipulation of military Command, Control, Communications, Computers, Intelligence, Surveillance and Reconnaissance (C4ISR) capabilities and the compromise of civilian support networks. Attacks on civilian infrastructure remain an option, and in the future attacks robotic military platforms are possible [29]. These tactics have a number of dependencies. First, the conflicting parties must have comparable military capabilities. Disrupting an opponent's C4ISR will be of little use if they still have the military superiority needed to achieve their objectives. Second, one state's disruption or destruction of another's cyber-infrastructure is only effective if they can defend their own assets, or have the capability to act without these assets with a minimal degradation in operational effectiveness. Third, states must have the resources required to deploy cyberweapons, which increase commensurate with the effectiveness of OCWC. Finally, cyberweapons usually rely on aggressive forward reconnaissance into networks of potential adversaries – the weapons should be positioned before conflict begins, which creates risks if an opponent discovers and traces a dormant cyberweapon. A further risk is the unpredictability of OCWC. Once unleashed the course of these weapons may 'be hard to predict, control, or contain' [14]. Unforeseen results may undermine relationships [22] or spread to unrelated states who then take retaliatory action.

Small state use of OCWC for coercion is similar to using them against state infrastructure in a warfighting scenario. It has the same dependencies regarding the relative size and cyber-dependence of an opponent, and shares the same risk regarding weapons acting in unforeseen ways. From a practical perspective, OCWC use for deterrence is little different from OCWC use for coercion. Both uses rely on the same aggressive forward reconnaissance of a potential opponent's network, so the difference between them becomes a matter of intent, which is difficult to prove. Another potential use of OCWC is defense diplomacy, which focuses on providing forces 'to dispel hostility, build and maintain trust and assist in the development of ... armed forces' [30]. Activities include training, bilateral and multilateral personnel exchanges, and joint military exercises. This could be expanded to encompass cyber-exercises conducted by military cyber-specialists. Defence diplomacy can act as a deterrent, but is only effective if relevant military capabilities are credible [31].

Having identified potential uses of OCWC, it becomes necessary to also identify how they may advance the political objectives of small states. A small state's political objectives depend on its behavior and identity, both internally and in the international system, which can be hard to quantify. Some predicative analysis, however, is possible through the use of conceptual security models – especially if analysis is completed across multiple models. Burton's [1] literature synthesis argues that small state security policy can be grouped under the conceptual models of *alliances, institutional cooperation*, and *identity and norms*. An alternative model emphasizes the two policy options of *collaborative influence* or *defensive autonomy* [25]. Synthesis of these approaches creates four models: alliances and collaborative influence, international cooperation and collaborative influence, identity and norms and collaborative influence, and identity and norms and defensive autonomy. Each model may have a greater or lesser amount of explanative power depending on the characteristics of the small state in question, but may provide substantive indicators for directional decisions.

The alliances and collaborative influence model presents small states with persuasive reasons to consider acquiring OCWC. This applies both to balancing behavior – joining an alliance against a threatening state, and bandwagoning, entering into an alliance with a threatening state [1]. The additional military resources provided by an alliance present greater opportunities for the exploitation of vulnerabilities caused by OCWC. In the event that a cyberweapon unwittingly targets a powerful third party, a small state may be less likely to be subjected to blowback if they are shielded by a strong alliance. Furthermore, OCWC may be a cost effective contribution to an alliance; a powerful state could even provide preferential OCWC procurement opportunities for a favored ally.

The institutional cooperation and collaborative influence model assumes that small states can exert influence by strengthening international organizations, encouraging cooperative approaches to security, and creating laws and norms to constrain powerful states [1]. Small states acting under this model will favor diplomatic and ideological methods of influence; and as such may be less likely to seek to acquire OCWC. Instead it is more probable that they will attempt to regulate OCWC in a manner similar to the restrictions on biological and chemical weapons, or by expanding current laws of international warfare to explicitly include cyberweapons.

As previously noted, the identity and norms model can be adapted to the pursuit of either collaborative-influence or defensive-autonomy. What is crucial to both variants of this model is the analysis of a small state's 'security identity', which grows from perceptions of 'past behavior and images and myths linked to it which have been internalized over long periods of time by the political elite and population of the state' [24]. This identity can be based around a number of disparate factors such as ongoing security threats, racial homogeneity, and parochialism. A state's security identity can lead it towards a collaborative security approach or a defensive, autonomous position, affecting the desirability of OCWC acquisition. It is this key divergence point combined with the above discussion that the authors present the acquisition framework in the next section.

4 Small State Acquisition of Offensive Cyberwarfare Capabilities: An Analytical Framework

The discussion and analysis offered in the previous sections suggests that a universally applicable recommendation on whether small states should acquire cyberweapons is not practical. The parameters of this decision depend too heavily on the behavior and identity of each particular small state and the scope of these attributes is too great to make universal pronouncements. Rather what is suggested is an analytical framework to

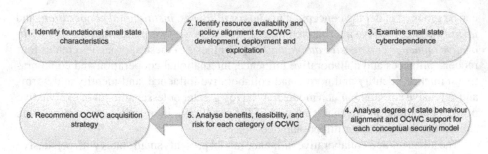

Fig. 1. OCWC acquisition framework

provide a customized evaluation of whether a particular small state should acquire OCWC. The first iteration of such a framework is provided below in Fig. 1.

Each step will be explained by a purpose statement, then demonstrated through a case study of New Zealand, chosen because it is both widely perceived as and self-identifies as a small state [32]. Ideally each step of the framework would be collaboratively completed by a group representing a variety of viewpoints from different government entities and academic specialties. There is the potential for a much more detailed evaluation than that presented, which has been condensed for brevity.

Step One – Identify foundational small state characteristics

Purpose: To identify key characteristics of the small state, under the categories of quantitative, qualitative-behavioral, and qualitative-identity. The starting point for determining these characteristics are the identifying measures that allow a state to be defined as 'small'.

Quantitative: New Zealand has a small population (approximately 4.5 million), a small GDP (approximately 197,000 million), and a small land area [33]. It is geographically isolated, bordering no other countries.

Qualitative-Behavioral: New Zealand practices an institutionally focused multilateral foreign policy [34]. It is a founding member of the United Nations and was elected to the Security Council for the 2015-2016 term after running on a platform of advocating for other small states. It participates in multiple alliances and takes a special interest in the security of the South Pacific [35].

Qualitative-Identity: New Zealand's self-identity emphasizes the values of fairness, independence, non-aggression, cooperation, and explicit acknowledgement of its status as a small state [32]. Its security identity is driven by a lack of perceived threat [36] that sometimes allows New Zealand to make security decisions based on principle rather than practicality. This was demonstrated by the banning of nuclear armed and powered ships within New Zealand waters, and its subsequent informal exclusion from aspects of the ANZUS Treaty. Despite reduced security, however, domestic opinion strongly supported the anti-nuclear policy [37], which, along with support for non-proliferation and disarmament, has strengthened the pacifistic elements of New Zealand's national identity.

Step Two – Identify resource availability and policy alignment for OCWC development, deployment, and exploitation

Purpose: To identify: how OCWC use aligns with current security and defense policies; whether the small state has the military capabilities to exploit vulnerabilities caused by OCWC deployment (against both military and civilian targets); and whether the small state has the intelligence and technical resources needed to target, develop and deploy OCWC.

In key New Zealand defense documents [35, 36, 38], mentions to the cyber-domain primarily refer to defense against cyber-attacks, with only two references to the application of military force to cyberspace, and no mention of acquisition of OCWC. New Zealand's defense policy has focused on military contributions to a safe and secure New Zealand, a rules-based international order, and a sound global economy. As the likelihood of direct threats against NZ and its closest allies is low, there has been a focus on peacekeeping, interoperability, disaster relief, affordability, and maritime patrol. New Zealand's military is numerically small (11,500 personnel including reservists) with low funding (1.1 % of GDP). Military capabilities include deployable ground forces, logistic capabilities, C4ISR, and limited but credible combat capabilities [35]. Because of its current lack of offensive focus, the New Zealand military lacks the ability to exploit vulnerabilities caused by the successful use of OCWC.

New Zealand is a member of the 'Five Eyes' intelligence network, and as such has access to a much greater range of intelligence [1] than most small states, which can be used to increase its ability to target and deploy OCWC. It has a modern Signals Intelligence (SIGINT) capability, housed by the civilian Government Communications Security Bureau (GSCB), which also has responsibility for national cybersecurity. It most likely has the technical capability to adapt existing cyberweapons or develop new ones, particularly if aided by its allies. Due to fiscal constraints, however, any additional funding for OCWC will most likely have to come from the existing defense budget [35] and thus result in compromises to other capabilities.

Step Three – Examine small state cyberdependence

Purpose: To examine the small state's reliance on cyberspace for its military capabilities and critical infrastructure, and its relative cyberdependence when compared to potential military opponents.

New Zealand has moderate to high cyberdependence, with increasing reliance on online services and platforms by individuals, organizations, military forces, and other government entities. This dependence will likely increase over the next few years. For example, the acquisition of additional C4ISR capabilities and initiatives to increase military adoption of netcentric warfare principles [36] will create new vulnerabilities. New Zealand's cyberdependence is further increased by limited cybersecurity expertise [1]. New Zealand does not have obvious military opponents so its relative level of cyberdependence is difficult to calculate.

<u>Step Four – Analyze degree of state behavior alignment and OCWC support for each conceptual security model</u>

Purpose: To identify the extent to which the small state's behavior aligns with each security model (high, medium, or low) and the extent to which OCWC would support or detract from the effectiveness of state behavior under each security model.

Alliances and Collaborative Influence: New Zealand maintains a close military alliance with Australia and is a member of the Five Power Defence Arrangement. New Zealand has also recently signed cybersecurity agreements with NATO and the UK [1]. The alliances above have focused on security and mutual defense, rather than offensive capabilities. New Zealand does however, have a policy of complementing Australian defense capabilities [36]. This could be achieved through the acquisition of OCWC, so long as this was closely coordinated and integrated with the Australian military. *State Behavioral Alignment: Medium/High*

International Cooperation and Collaborative Influence: New Zealand usually pursues a multilateral foreign policy approach and is a member of multiple international organizations. It has a long history of championing disarmament and arms control [34], which conflicts with the acquisition of new categories of offensive weapons. *State Behavioral Alignment: High*

Identity and Norms and Collaborative Influence: With regard to collaboration, New Zealand's identity and norms strike a balance between practicality and principle. It wishes to advance what it regards as important values, such as human rights and the rule of law [32]. It however, still wishes to work in a constructive and practical manner. Procurement of OCWC is unlikely to advance this model. *State Behavioral Alignment: Medium*

Identity and Norms and Defensive Autonomy: Despite its multilateral behavior, NZ takes pride in maintaining independent views on major issues [32]. Its isolation and a lack of major threats has allowed it to retain a measure of autonomy in its defense policy and maintain a small military. Its independent and pacifistic nature suggest that OCWC acquisition could be controversial. *State Behavioral Alignment: Low/Medium*

<u>Step Five – Analyze benefits, feasibility and risk for each category of OCWC</u>

Purpose: To identify the benefits, feasibility, and risk of acquiring OCWC based on each category of potential use, as shown in Fig. 2, then to analyze this information against the degree of OCWC support for different security models in step four, as shown in Fig. 3. This results in a ranking of the benefits, feasibility, and risk under each combination of OCWC use and small state security model, as well as an overall recommendation for OCWC acquisition under each security model and OCWC use.

	Warfighting	Coercion	Deterrence	Defense Diplomacy
Benefits	Ability to complement military capabilities of allies Cost effective offensive capability	Limited coercive ability from OCWC	Limited deterrence from OCWC	Deterrence from demonstrating effective OCWC via defence diplomacy
Feasibility	Allies may provide favourable procurement opportunities Appropriate technical and intelligence resources exist	Appropriate technical and intelligence resources exist	Appropriate technical and intelligence resources exist	Appropriate technical and intelligence resources exist
Risks	Procurement may result in reduced funding for other military capabilities Domestic opposition to acquisition of new offensive weapons OCWC acquisition may reduce international reputation OCWC exploitation relies on allied forces High level of cyberdependence increases vulnerability to retaliation	Domestic opposition to acquisition of new offensive weapons Security identity not reconcilable with coercive military actions Procurement may result in reduced funding for other military capabilities OCWC acquisition may reduce international reputation High level of cyberdependence increases vulnerability to retaliation	Procurement may result in reduced funding for other military capabilities OCWC acquisition may reduce international reputation High level of cyberdependence increases vulnerability to retaliation Lack of identified threats reduces ability to target and develop deterrent OCWC	Procurement may result in reduced funding for other military capabilities OCWC acquisition may reduce international reputation High level of cyberdependence reduces deterrent effect

Fig. 2. OCWC benefits, feasibility and risk matrix

The authors accept that the rankings above are subjective in nature and require additional subject matter expertise and collaborative methods before they are relied upon by policy officials. They are offered in this spirit.

<u>Step Six – Recommended OCWC acquisition strategy</u>

Purpose: To summarise key findings, recommend if a small state should acquire OCWC, and identify the next steps advising how the framework's output should be used.

OCWC Acquisition Matrix: New Zealand						
Security Model	BFR	Warfighting	Coercion	Deterrence	Defense Diplomacy	Overall
Alliances and collaborative influence	Benefits	Medium	Low	Low	Medium	**Medium**
	Feasibility	Medium	Medium	Medium	Medium	**Medium**
	Risks	High	Very High	High	Low	**High**
	Recommendation	**Further Investigation**	**No**	**No**	**Further Investigation**	**Further Investigation**
International cooperation and collaborative influence	Benefits	Low	Low	Low	Medium	**Low**
	Feasibility	Medium	Medium	Medium	Medium	**Medium**
	Risks	High	High	High	Low	**High**
	Recommendation	**No**	**No**	**No**	**Further Investigation**	**No**
Identity and norms and collaborative influence	Benefits	Low	Low	Low	Medium	**Low**
	Feasibility	Medium	Medium	Medium	Medium	**Medium**
	Risks	High	High	High	Low	**High**
	Recommendation	**No**	**No**	**No**	**Further Investigation**	**No**
Identify and norms and defensive autonomy	Benefits	Low	Low	Low	Low	**Low**
	Feasibility	Medium	Medium	Medium	Medium	**Medium**
	Risks	High	High	High	Low	**Low**
	Recommendation	**No**	**No**	**No**	**No**	**No**

Fig. 3. OCWC acquisition matrix: New Zealand

Key Findings: New Zealand is unlikely to reap significant benefits from the acquisition of OCWC. This is due to its limited military capabilities, multilateral foreign approach, extensive participation in international organizations, and pacifistic security identity. The most likely factors to change this evaluation and increase the benefits of OCWC acquisition would be an increased focus on military alliances, the emergence of more obvious threats to New Zealand sovereignty, and a changing security identity.

Recommendation: It is recommended that New Zealand **does not acquire OCWC** at this time.

Next Steps: The output of this framework can be incorporated into relevant defense capability and policy documents. *If the framework had recommended the acquisition of OCWC*, then its output could be used to inform specific strategic, operational, and thus system requirements for OCWC. These capabilities could then be analysed under a standard return on investment business case model, in which a more detailed analysis of benefits, costs, and risks would allow an appropriate course of action to be decided in a transparent and fiscally responsible manner.

5   Conclusions

Recent analysis of cyberwarfare has been dominated by works focused on waning American hegemony, a rising China, Russian revanchism, and, growing cyber-bellicosity from rogue states and non-state actors. While not questioning the importance of these geopolitical trends, this paper shifts analysis to a relatively unexplored area – the factors that motivate a small state to acquire OCWC and the conditions under which acquisition would be beneficial. It offers a definition of cyberwarfare that focuses on its political and kinetic nature, complemented by analysis that challenges overestimation of OCWC. This is achieved through an exploration of the limitations of OCWC and the concept of cyberpower, arguing that OCWC can neither win military conflict unaided, nor alter principles of warfare. Second, it analyses both theoretical and likely uses of OCWC by small states. It argues that definitional tensions regarding the term 'small state' can be resolved by a definition that relies on the overlapping, qualitative and quantitative properties that are demonstrative of its identity and behavior. The paper then turns analyzes four categories of potential OCWC use, which are examined with regard to their potential to advance small state political objectives, as explained via multiple conceptual small state security models.

Having concluded that a universally applicable recommendation on whether small states should acquire cyberweapons is not possible, this paper instead presents an analytic framework designed to produce individualized recommendations on whether a particular small state should acquire OCWC. The framework has been demonstrated by a case study on a quintessential small state – New Zealand. It began with an analysis of the quantitative and qualitative characteristics that could be used to identify New Zealand as a 'small' state, followed by an examination of its security and defense policies, military capabilities, and technical, financial and intelligence resources. This was augmented by consideration of the extent of New Zealand's 'cyber-dependence' and a behavioral analysis of New Zealand and its potential uses of OCWC under small state security models. The results of this analysis have been assessed against each category of potential OCWC use, resulting in predictive information regarding the utility of OCWC and the overall desirability of OCWC acquisition. The framework demonstrates that New Zealand, with its limited military capabilities, absence of direct threats, institutionally focused foreign policy, and pacifistic security identity, is unlikely to benefit from the acquisition of OCWC at this time. This result, however, is unique to New Zealand; further small state examinations will enhance the OCWC acquisition framework offered as well as its utility in this decision process. The spectrum of small state behavior and identity is far-reaching; each small state must examine its own circumstances to determine whether the acquisition of OCWC will allow it to advance its own national security interests.

References

1. Burton, J.: Small states and cyber security: the case of New Zealand. Polit. Sci. (00323187), **65**(2), 216–238 (2013). doi:10.1177/0032318713508491
2. Mayer, M., Carpes, M., Knoblich, R.: (Introduction) The Global Politics of Science and Technology. Springer, Berlin (2014)

3. Vice Chairman of the Joint Chiefs of Staff: Joint Terminology for Cyberspace Operations (2011). http://www.nsci-va.org/CyberReferenceLib/2010-11-joint%20Terminology%20for%20Cyberspace%20Operations.pdf
4. Schrier, F.: On Cyberwarfare DCAF Working Paper No. 7 (2015). www.dcaf.ch/content/download/67316/.../OnCyberwarfare-Schreier.pdf
5. US Department of Defense. The DOD Cyberstrategy (2015). http://www.defense.gov/home/features/2015/0415_cyber-strategy/Final_2015_DoD_CYBER_STRATEGY_for_web.pdf
6. Shakarian, P., Shakarian, J., Ruef, A.: Introduction to Cyberwarfare: A Multidisciplinary Approach. Syngress, Burlington (2013)
7. Theohary, C., Rollins, J.: Cyberwarfare and Cyberterrorism: In Brief. Congressional Research Service, Washington (2015). https://www.fas.org/sgp/crs/natsec/R43955.pdf
8. Colarik, A., Janczewski, L.: Developing a grand strategy for Cyber War. In: 2011 The International Conference on Information Assurance & Security (IAS), pp. 52–57 (2011). doi: 10.1109/ISIAS.2011.6122794
9. Parks, R.C., Duggan, D.P.: Principles of Cyberwarfare. IEEE Secur. Priv. Mag. 9(5), 30 (2011). doi:10.1109/MSP.2011.138
10. Arquilla, J.: Twenty years of Cyberwar. J. Mil. Ethics 12(1), 80–87 (2013)
11. Kiravuo, T., Tiilikanien, S., Sarela, M., Manner, J.: Peeking under the skirts of a nation: finding ics vulnerabilities in the critical digital infrastructure. In: Proceedings of the European Conference on e-Learning (2015)
12. Korns, S., Kastenburg, J.E.: Georgia's Cyber left hook. Parameters 38(4), 60–76 (2009). Winter 08-09
13. Singer, P.W., Friedman, A.: Cybersecurity and Cyberwar: What Everyone Needs to Know. Oxford University Press, Oxford (2014)
14. Gompert, D. C., Libicki, M.: Waging cyber war the american way. Survival (00396338), 57(4), 7–28 (2015). doi:10.1080/00396338.2015.1068551
15. Mahnken, T.: Cyberwar and Cyberwarfare. America's Cyber. Future 2, 53–62 (2011). https://www.google.com.tw/webhp?sourceid=chrome-instant&ion=1&espv=2&ie=UTF8#q=Chapter+iV:+Cyberwar+and+Cyber+Warfare+By+Thomas+G.+Mahnken+citation
16. Rid, T., McBurney, P.: Cyber-Weapons. RUSI J. R. U. Serv. Inst. Def. Stud. 157(1), 6 (2012). doi:10.1080/03071847.2012.664354
17. Carr, J.: The misunderstood acronym: why cyber weapons aren't WMD. Bull. At. Sci. 69(5), 32 (2013). doi:10.1177/0096340213501373
18. Lewis, J., Timlin, K.: Cybersecurity and Cyberwarfare 2011 preliminary assessment of national doctrine and organization (2011). http://unidir.org/files/publications/pdfs/cyber security-and-cyberwarfare-preliminary-assessment-of-national-doctrine-and-organization-380.pdf
19. Gartzke, E.: The myth of cyberwar: bringing war in cyberspace back down to earth. Int. Secur. 2, 41 (2013)
20. Rid, T.: Cyberwar and Peace. Foreign Aff. 92(6), 77–87 (2013)
21. Clarke, R.D., Knake, R.K.: Cyber War. HarperCollins, New York (2010)
22. Nye, J.: Cyber Power (2010). http://belfercenter.ksg.harvard.edu/files/cyber-power.pdf
23. Kuehl, D.T.: From Cyberspace to Cyberpower: Defining the Problem. Cyberpower and National Security, Washington (2009). http://ctnsp.dodlive.mil/files/2014/03/Cyberpower-I-Chap-02.pdf
24. Goetschel, L.: The foreign and security policy interests of small states in today's Europe. In: Goetschel, L. (ed.) Small States Inside and Outside the European Union, pp. 13–31. Kluwer Academic Publishers, Dordrecht (1998)

25. Rickli, J.: European small states' military policies after the Cold War: from territorial to niche strategies. Camb. Rev. Int. Aff. **21**(3), 307–325 (2008). doi:10.1080/09557570802253435
26. Sutton, P.: The concept of small states in the international political economy. Round Table **100**(413), 141–153 (2011)
27. Baehr, P.: Small states: a tool for analysis? World Polit. **27**, 456–466 (1975). doi: 10.2307/2010129
28. Wittgenstein, L., Anscombe, G.M.: Philosophical Investigations. The United Kingdom Basil Blackwell, Oxford (1958). c1953
29. Schutte, S.: Cooperation beats deterrence in Cyberwar. Peace Econ. Peace Sci. Public Policy **18**(3), 1–11 (2012). doi:10.1515/peps-2012-0006
30. Ministry of Defence Policy Papers Defence Diplomacy (1998). http://webarchive.nationalarchives.gov.uk/20121026065214/http://www.mod.uk/NR/rdonlyres/BB03F0E7-1F85-4E7B-B7EB-4F0418152932/0/polpaper1_def_dip.pdf
31. Tan, A.T.: Punching above its weight: singapore's armed forces and its contribution to foreign policy. Def. Stud. **11**(4), 672–697 (2011)
32. McLay, J.: New Zealand and the United Nations: Small State, Big Challenge, August 2013. http://www.nzunsc.govt.nz/docs/Jim-McLay-speech-Small-State-Big%20Challenge-Aug-13.pdf
33. Statistics New Zealand, (n.d.): Index of key New Zealand Statistics. http://www.stats.govt.nz/browse_for_stats/snapshots-of-nz/index-key-statistics.aspx#
34. Ministry of foreign affairs and trade. Foreign Relations, March 2014. http://mfat.govt.nz/Foreign-Relations/index.php
35. New Zealand Defence Force. Defence Capability Plan (2014). http://www.nzdf.mil.nz/downloads/pdf/public-docs/2014/2014-defence-capability-plan.pdf
36. New Zealand Defence Force. Defence White Paper 2010 (2010). http://www.nzdf.mil.nz/downloads/pdf/public-docs/2010/defence_white_paper_2010.pdf
37. Reitzig, A.: In defiance of nuclear deterrence: anti-nuclear New Zealand after two decades. Med. Conflict Surv. **22**(02), 132–144 (2006). doi:10.1080/13623690600621112
38. New Zealand Defence Force. New Zealand Defence Force Doctrine (2012). http://www.nzdf.mil.nz/downloads/pdf/public-docs/2012/nzddp_d_3rd_ed.pdf

Data-Driven Stealthy Injection Attacks on Smart Grid with Incomplete Measurements

Adnan Anwar[✉], Abdun Naser Mahmood, and Mark Pickering

School of Engineering and Information Technology, UNSW,
Canberra, ACT 2600, Australia
{adnan.anwar,M.Pickering}@adfa.edu.au, Abdun.Mahmood@unsw.edu.au

Abstract. Key smart grid operational module like state estimator is highly vulnerable to a class of data integrity attacks known as 'False Data Injection (FDI)'. Although most of the existing FDI attack construction strategies require the *knowledge* of the power system topology and electric parameters (e.g., line resistance and reactance), this paper proposes an alternative data-driven approach. We show that an attacker can construct stealthy attacks using only the subspace information of the measurement signals without requiring any prior power system *knowledge*. However, principle component analysis (PCA) or singular value decomposition (SVD) based attack construction techniques do not remain stealthy if measurement signals contain missing values. We demonstrate that even in that case an intelligent attacker is able to construct the stealthy FDI attacks using low-rank and sparse matrix approximation techniques. We illustrate an attack example using augmented lagrange multiplier (ALM) method approach. These attacks remain hidden in the existing bad data detection modules and affect the operation of the physical energy grid. IEEE benchmark test systems, different attack scenarios and state-of-the-art detection techniques are considered to validate the proposed claims.

Keywords: False injection · Smart grid · State estimator · Blind attack · PCA · SCADA · EMS

1 Introduction

Most of the recent cyber-attacks in Energy System target the operation centres to jeopardise the physical operation of the grid by cyber intrusions. One such possible threat is stealthy data integrity attacks that remains hidden in the existing anomaly detection modules and mislead the system operators by providing wrong system states [20], as discussed below.

Energy Management System (EMS) is a collection of software tools which is used by the operators to monitor, control, and optimize the performance of the energy system. EMS receives real-time measurement data from the SCADA (supervisory control and data acquisition) system and takes intelligent control

© Springer International Publishing Switzerland 2016
M. Chau et al. (Eds.): PAISI 2016, LNCS 9650, pp. 180–192, 2016.
DOI: 10.1007/978-3-319-31863-9_13

decisions using different operational modules. SCADA communication protocols have been proven vulnerable due to the *spoofing* or *man-in-the-middle* attacks [22]. Besides, cyber-attacker can compromise networked sensor devices and manipulate the original measurements. As a result, EMS receives false measurement information for decision making. In an EMS, Bad Data Detector (BDD) is used to identify the anomalous data based on statistical analysis and hypothesis testing [2,20]. This module can easily identify the attacked measurements if the attack vector is constructed randomly. This module has been widely used in the industrial energy system operational centres for the last decades without any concern. However, in a seminal work by Liu et al., it has been revealed that the existing BDD module fails to detect the presence of the injection attack if the attack vector is constructed strategically with some principles derived from power system topology and connectivity information [19]. This class of strategic data integrity attack is well known as 'False Data Injection Attack (FDI)' in the smart grid research.

At present, a good number of research works have been conducted to investigate different aspects of FDI attacks. Significant research works have been carried out on FDI attacks to investigate their stealthiness and construction strategies [6,7,10,11,13–15,20,21,26], along with prospective detection and prevention measures [3–5,8,11,12]. Most of the attack construction strategies mentioned above consider that the attacker has full or partial information of the system topology and electric parameters of the transmission lines. These all information can be obtained from the system jacobian (H matrix) that is used during the state estimation. If the attacker knows the H matrix information, then stealthy FDI can be constructed as demonstrated by Liu et al. [19,20]. Similar to [19,20], Sinopoli et al. considers complete knowledge of H matrix as well in [25]. In [23], authors propose a modified form of FDI attack which can be constructed based on partial *system knowledge*. The topology information can be obtained from an insider attack or by gaining illegitimate access of the database that contains topology and connectivity information. Although large number of research works in this stream consider that the attacker has gained the knowledge about the system topology and electric parameters, in practice it is very difficult to obtain such information.

Very recently, Yu et al. [26] and Kim et al. [14] have demonstrated a *blind* attack construction strategy that do not rely on any prior power system topological and electric line parameter information. In those data-driven approaches, the attack vector is constructed based on the subspace information of the measurement signals only. Yu et al. [26] proposes a principle component analysis (PCA) and Kim et al. [14] proposes a singular value decomposition (SVD) based subspace estimation technique for stealthy attack construction. Actually, power system topology and electric parameter can be considered static (fixed) and the system dynamics change very slowly (due to load changes which are independent from each other) for a small amount of time under normal operating condition. Therefore, equivalent knowledge required for stealthy attack construction can be approximated from the variations of the measurement signals only (See Sect. 4

for details). This idea is explored in [14,26] in detail, where authors show that this data-driven attack remains hidden in the existing BDD module.

In this work, first we implement the PCA based data-driven FDI attack. Next we show that the PCA based attack strategy fails to remain stealthy in the existing BDD module if the measurement signals contain any missing value. In those cases, subspace estimation using PCA is incorrect and attack based on that wrong information can not guarantee its stealthiness. In SCADA communication, missing value is very common which may occur due to communication or device failure. Although PCA based approach can not be used to construct stealthy attacks, in our research we have seen that an intelligent attacker can circumvent this problem by utilizing an alternative form of attack. We assume that missing measurements will be only a small fraction of all measurements, hence they are sparse in nature. On the other hand, gradually changing system states will lead to a low-rank measurement matrix [18]. Hence, an intelligent attacker can utilize a low-rank and sparse matrix factorization technique on the original measurement matrix with missing values. In this work we use augmented lagrange multiplier (ALM) method to approximate the original measurement matrix by estimating the missing values. Next,we show that a stealthy attack can be constructed based on the estimated low-rank measurement matrix. We have considered IEEE benchmark test systems to validate the proposed attack strategy. The traditional as well as recently proposed *blind* FDI attack construction strategies and no-attack cases are also explored to compare with the proposed method.

2   Measurement Model

Energy Management System (EMS) is a set of software programs which can take intelligent operational decisions. State Estimators and Bad Data Detectors (BDD) are two vital operational modules of EMS. Typically, EMS receives measurement data from the supervisory control and data acquisition (SCADA) system. State estimator (SE) and BDD are responsible for the treatment of measurement data that includes noise suppression, missing data handling and identification of anomalous data. The state estimation problem is formulated based on the principle of a *weighted least square (WLS)* estimator as follow:

$$\underset{y}{\operatorname{argmin}} J(\mathbf{y}) = \frac{1}{\sigma^2} \|\mathbf{z} - \mathbf{Hy}\|_2^2 \tag{1}$$

where $\mathbf{z}$ is the measurement vector, $\mathbf{y}$ is the power system state vector and $\mathbf{H}$ is the system Jacobian matrix. Under this setup, the residual becomes,

$$\mathbf{r} = \mathbf{z} - \mathbf{Hy} \tag{2}$$

If there are m measurement devices and n system states, then $\mathbf{H}$ is an $m \times n$ matrix. In order to estimate the state vector accurately, system observability must be maintained. A system will be *observable* if H is a full rank matrix which leads to the assumptions that $m \geq n$ and $rank(H) = n$ [14].

The chi-square (χ^2) test is widely used to detect bad measurement data [2]. As it is assumed that noise samples follow a *normal* distribution with zero mean and they are independent, $J(\mathbf{y})$ will follow a (χ^2) distribution with a ψ degree of freedom, where $(\psi = m - n)$ [2]. Considering a desired significance level (e.g., 95 %), a threshold- $\chi^2_{(m-n),p}$ from the chi-square distribution can be obtained. If there exists an anomalous (bad) data sample, the value of $J(\mathbf{y}) \geq \chi^2_{(m-n),p}$.

3   Attack Based on Known System Jacobian

During False Data Injection (FDI) attack, attack vector $\mathbf{a}$ is injected with the original measurements $\mathbf{z}$ which leads to the new manipulated measurement vector $\mathbf{z_a}$, where $\mathbf{z_a} = \mathbf{z} + \mathbf{a}$. Due to this false data injection, let us assume that the original state vector $\mathbf{y}$ deviates by a vector $\mathbf{c}$ and new state vector becomes $\mathbf{y_a}$. So, $\mathbf{y_a} = \mathbf{y} + \mathbf{c}$. Liu et al. [19] are the first to observe that the attack will be hidden in the existing BDD module if the attack vector is constructed following $\mathbf{a} = \mathbf{Hc}$, where $\mathbf{c}$ can be chosen randomly or strategically based on how much deviation the attacker wants to cause to the system states. For any value of $\mathbf{c}$, the residual obtained after the attack (r_{attack}) would be the same with the residual when there is no injection attack (r_{normal}). The proof is given below [20]:

$$r_{attack} = \|\mathbf{z}_a - \mathbf{Hy}_a\|$$
$$= \|\mathbf{z} + \mathbf{a} - \mathbf{H}(\mathbf{y} + \mathbf{c})\|$$
$$= \|\mathbf{z} + \mathbf{a} - \mathbf{Hy} - \mathbf{Hc}\|$$
$$= \|\mathbf{z} - \mathbf{Hy}\| \quad (as, \ \mathbf{a} = \mathbf{Hc})$$
$$= r_{normal}$$

As both of the residuals are the same, the detection probability of the attacked case will follow a similar trend to the attack-free normal case. Hence, the residual of the cyber attack will be less than the threshold and attack will remain hidden in the existing detection modules.

4   Stealthy Attack without System Jacobian

The stealthy attack strategy described in the above section requires the information of the system Jacobian ($\mathbf{H}$) matrix. In practice it is very difficult to obtain the system topological connectivity and line admittance information (represented as $\mathbf{H}$ matrix). In a recent study [26], Yu et al. has shown that stealthy attack can be constructed using only the measurement signals without requiring the system Jacobian ($\mathbf{H}$) matrix. A brief discussion on how to construct the attack vector is given below.

Consider the time-series measurement matrix $\mathbf{Z}_{d \times m}$, where each row represents a time instant and each column corresponds to the measurement variables. Now, PCA is applied on the data matrix $\mathbf{Z}_{d \times m}$. PCA is a multivariate statistical technique which can transform the correlated observations into uncorrelated variables known as principal components. These orthogonal principal components are the linear combinations of the original observations. After a successful PCA transformation, we obtain the vector of principal components $\mathbf{x}$ and the transformation matrix $\tilde{\mathbf{M}}$ [26]. Therefore, the PCA relationship can be represented as:

$$\tilde{\mathbf{M}}^T \mathbf{Z} = \mathbf{x} \tag{3}$$

Now, the measurement matrix $\mathbf{Z}$ can be approximated by,

$$\mathbf{Z} \approx \begin{bmatrix} \tilde{M}_{1,1} & \tilde{M}_{1,2} & \cdots & \tilde{M}_{1,m} \\ \vdots & \vdots & \ddots & \vdots \\ \tilde{M}_{n,1} & \tilde{M}_{n,2} & \cdots & \tilde{M}_{n,m} \\ \vdots & \vdots & \ddots & \vdots \\ \tilde{M}_{m,1} & \tilde{M}_{m,2} & \cdots & \tilde{M}_{m,m} \end{bmatrix} \begin{bmatrix} x_1 \\ \vdots \\ x_n \\ \vdots \\ x_m \end{bmatrix} \tag{4}$$

where the principal components and the corresponding eigenvectors are arranged based on their eigenvalues in descending order. As the system jacobian $\mathbf{H}$ is an n rank matrix, only n eigenvectors are considered. So, (4) can be summarized as:

$$\mathbf{Z} \approx \begin{bmatrix} \tilde{M}_{1,1} & \tilde{M}_{1,2} & \cdots & \tilde{M}_{1,n} \\ \vdots & \vdots & \ddots & \vdots \\ \tilde{M}_{m,1} & \tilde{M}_{m,2} & \cdots & \tilde{M}_{m,n} \end{bmatrix} \begin{bmatrix} x_1 \\ \vdots \\ x_n \end{bmatrix} \tag{5}$$

$$\equiv \mathbf{H}_{pca} \mathbf{x}_{pca} \tag{6}$$

where the reduced transformation matrix $\tilde{\mathbf{M}}_{m \times n}$ has been considered as $\mathbf{H}_{pca}$ which is then used for stealthy attack construction. In that case, the attack vector will be,

$$\mathbf{a}_{pca} = \mathbf{H}_{pca} \mathbf{c} \tag{7}$$

where $\mathbf{c}$ is an arbitrary non-zero vector of length n. Therefore, the attacked measurement vector becomes $\mathbf{z}_{pca} = \mathbf{z} + \mathbf{a}_{pca}$. Further details and proof of stealthiness can be obtained from [26].

5   Attack with Incomplete Measurements

The above discussed *blind* attack strategy is constructed only from the measurement signals without requiring system Jacobian ($\mathbf{H}$) matrix. However, one major challenge is to handle *missing* measurements which is common in a real-world electric grid. Measurements can be missing due to communication loss or device unavailability or malfunction. Therefore, in a practical case it is more logical to assume that the observed measurement matrix is incomplete. That means, some measurements are not available in the observed measurement matrix. However, in a practical scenario, the number of missing measurements is very less compared to the total number of measurements. Hence, missing measurements are *sparse* in nature. Now, the construction of stealthy FDI attacks using the PCA based method discussed above is not possible in the presence of *missing* measurements. Even if there exists a single missing value in the measurement (or observation) matrix, the traditional PCA technique shows its brittleness [9]. In that case, the estimation of the coefficient matrix using PCA is not accurate and it makes the FDI attack detectable. In this work, we warn the system operator by demonstrating that intelligent attacker can still circumvent this challenge by applying an alternative form of stealthy FDI attack, as discussed below.

The original measurement matrix is a low rank matrix [18] and the missing values can be assumed to be sparse. Therefore, sparse optimization technique can be used to approximate the original low-rank measurement matrix from the observed measurement matrix (with incomplete information). If the original low-rank measurement matrix is $\mathbf{A}$, sparse matrix of missing values is $\mathbf{E}$, and observed measurement matrix with incomplete information is $\mathbf{Z}$, then we can form the relationship as below:

$$\mathbf{Z} = \mathbf{A} + \mathbf{E} \tag{8}$$

Now, we can formulate the problem as a *matrix recovery* problem and the exact recovery of $\mathbf{A}$ and sparse $\mathbf{E}$ can be represented mathematically as below:

$$min \ \|\mathbf{A}\|_* + \lambda \|\mathbf{E}\|_1, \quad s.t. \ \mathbf{Z} = \mathbf{A} + \mathbf{E} \tag{9}$$

In this convex optimization problem, $\|.\|_*$ and $\|.\|_1$ denotes the nuclear norm and l_1 norm of a matrix, respectively, and λ is a positive weighting parameter [17]. To solve this problem, we use augmented lagrange multiplier (ALM) [16] method as discussed below:

The ALM method can be used for the general constraint optimization problem as follows:

$$min \ f(x), \quad s.t. \ h(x) = 0 \tag{10}$$

Using the ALM method, the objective function of the above optimization problem can be written as a lagrangian function:

$$L(x, Y, \mu) = f(x) + \langle \gamma, h(x) \rangle + \frac{\mu}{2} \|h(x)\|_F^2 \tag{11}$$

where γ is the *lagrange multiplier* and μ is a positive scalar. Considering $x = (\mathbf{A}, \mathbf{E})$, $f(x) = \|\mathbf{A}\|_* + \lambda \|\mathbf{E}\|_1$ *and* $h(x) = \mathbf{Z} - \mathbf{A} - \mathbf{E}$, the Lagrangian function can be written as:

$$L(\mathbf{A}, \mathbf{E}, \gamma, \mu) = f(x) = \|\mathbf{A}\|_* + \lambda \|\mathbf{E}\|_1 + \langle \gamma, (\mathbf{Z} - \mathbf{A} - \mathbf{E}) \rangle + \frac{\mu}{2} \|\mathbf{Z} - \mathbf{A} - \mathbf{E}\|_F^2 \tag{12}$$

The solution optimization process is driven by the following two update steps,

$$\mathbf{A}_{k+1} = \arg \ \min \ L(\mathbf{A}, \mathbf{E}_k, \gamma_k, \mu_k) \tag{13}$$

$$\mathbf{E}_{k+1} = \arg \ \min \ L(\mathbf{A}_{k+1}, \mathbf{E}, \gamma_k, \mu_k) \tag{14}$$

Equation (13) can be computed from the soft-shrinkage formula [18], using an iterative thresholding (IT) approach that uses the singular value decomposition

(SVD) of the matrix $(\mathbf{Z} - \mathbf{E}_k + \mu_k^{-1}\gamma_k)$ [16]. After performing the SVD, the unitary matrix $\mathbf{U}$, $\mathbf{V}$ and the rectangular diagonal matrix $\mathbf{S}$ is obtained. Then, $\mathbf{A}$ is updated as,

$$\mathbf{A}_{k+1} = \mathbf{U}\xi_{\mu_k^{-1}}[\mathbf{S}]\mathbf{V}^T \tag{15}$$

and $\mathbf{E}$ is updated as,

$$\mathbf{E}_{k+1} = \xi_{\lambda\mu_k^{-1}}[\mathbf{Z} - \mathbf{A}_{k+1} + \mu_k^{-1}\lambda_k] \tag{16}$$

here $\lambda = 1/\sqrt{max(m,t)}$ and ξ is a soft-thresholding (shrinkage) operator, defined as [16]:

$$\xi_\varepsilon[x] = \begin{cases} x - \varepsilon, & if \ x > \varepsilon \\ x + \varepsilon, & if \ x < -\varepsilon \\ 0, & otherwise, \end{cases} \tag{17}$$

During each of the iterations, γ and μ are updated as follows:

$$\gamma_{k+1} = \gamma_k + \mu_k(\mathbf{Z} - \mathbf{A}_{k+1} - \mathbf{E}_{k+1}) \tag{18}$$

$$\mu_{k+1} = \Omega\mu_k \tag{19}$$

where Ω is a positive constant. The optimization process continues until the convergence criteria is satisfied. The convergence is checked based on the relative error using (20) against a tolerance, τ.

$$c_k^{idx} = \|\mathbf{Z} - \mathbf{A}_{k+1} - \mathbf{E}_{k+1}\|_F / \|\mathbf{Z}\|_F; \tag{20}$$

The proof of convergence of the ALM algorithm can be obtained from [16]. Once the approximation of the original low-rank measurement matrix is performed, then the attack is constructed using the that matrix (with no missing value) following the procedure discussed in Sect. 4. The complete procedure is summarized in Algorithm 1.

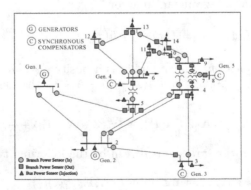

Fig. 1. Location and types of power measurement sensors for an IEEE 14 bus test system.

Algorithm 1. ALM based FDI attack strategy

---

1 Step 1: Deals with estimating the original measurements from the measurement matrix with missing values

---

 Input : Measurement matrix with missing values $\mathbf{Z}$, λ

2 Initialize $\mathbf{A_0}, \mathbf{E_0}, t_0, \mu_0, \bar{\mu}$;

3 **while** *not converged* **do**

4   // Solve $\mathbf{A}_{k+1} = \arg\ \min\ L(\mathbf{A}, \mathbf{E}_k, \gamma_k, \mu_k)$;

5      $\mathbf{A}_{k+1} = \mathbf{U}\xi_{\mu_k^{-1}}[\mathbf{S}]\mathbf{V}^T$;

6   // Solve $\mathbf{E}_{k+1} = \arg\ \min\ L(\mathbf{A}_{k+1}, \mathbf{E}, \gamma_k, \mu_k)$;

7      $\mathbf{E}_{k+1} = \xi_{\lambda\mu_k^{-1}}[\mathbf{Z} - \mathbf{A}_{k+1} + \mu_k^{-1}\lambda_k]$;

8   // Update γ and μ ;

9      $\gamma_{k+1} = \gamma_k + \mu_k(\mathbf{Z} - \mathbf{A}_{k+1} - \mathbf{E}_{k+1})$;

10      $\mu_{k+1} = \Omega\mu_k$;

11      $k \longleftarrow k+1$;

12 **end**

 Output: $A \longleftarrow A_k, E \longleftarrow E_k$

---

13 Step 2: Deals with constructing stealthy attack vector from the estimated measurements

---

14 $[\tilde{\mathbf{M}}]$=pca($\mathbf{A}$) // calculate coefficient matrix $\tilde{\mathbf{M}}$ using principal component analysis ;

15 Consider n eigenvectors of $\tilde{\mathbf{M}}$ and obtain $\mathbf{H}_{pca}$;

16 Generate a non-zero vector $\mathbf{c}$ of the length n ;

17 Construct attack vector, $\mathbf{a}_{pca} = \mathbf{H}_{pca}\mathbf{c}$;

18 Calculate manipulated sensor measurements, $\mathbf{z}_{pca} = \mathbf{z} + \mathbf{a}_{pca}$;

 Output: Stealthy corrupted measurements $\mathbf{z}_{pca}$

---

6 Results and Discussion

All experiments were conducted using IEEE benchmark 14 bus system [1,27]. IEEE 14 bus test system has 20 line sections (branches) and 14 buses (nodes). Therefore, the total measurements consist of 20 incoming power flow sensors, 20 outgoing power flow sensors, and 14 power injection sensors (total 54 sensors). These sensors are shown in Fig. 1, where the sensors are marked with individual symbols. There is no known publicly available real-world smart grid cyber-attack data [24]. Hence, realistic power system simulation is carried out using Matlab based simulation tool MATPOWER [27]. MATPOWER is widely used for simulating power system data and reflects a realistic simulation environment for the real-life complex power systems [10,26,27]. The state estimation formulation and attack construction strategies are also implemented in Matlab on a PC with an Intel(R) core i7 @ 3.4 GHz- 3.4 GHz processor and 16 GB of RAM.

Under normal operating condition (without any attack) following a chi-square test considering 95 % confidence interval, we obtain the threshold for BDD is 56.94 (as there are 54 measurements and 13 system states, the degrees of freedom are 41) [2]. First, we consider the *Ideal* and the *practical* case. By mentioning *Ideal* case, we refer to the scenario where there is no Gaussian noise in the measurement model and *practical* case includes i.i.d. (independent and identically distributed) gaussian noises of 20–35 db. Under this test setup we perform state estimation and BDD. The estimated measurements and the observed ones are plotted in Fig. 2 (a) and (b) for practical and ideal case, respectively. We see that for both cases, the estimated ones follow the observed ones. The residuals for these two scenarios are 39.069 and 8.67^{-26} respectively.

Now we inject known $\mathbf{H}$ based attack vector (discussed in Sect. 3) with the original measurements under practical scenario. The original and observed (attacked) measurements are represented with green and red symbols respectively in Fig. 2(c). Note that the system operators will receive the SCADA measurements without knowing whether the measurements are attacked or not. They will verify it by performing BDD operation. We observe that the estimated measurements (represented by the blue line) follow the attacked measurements (red) rather than the original. For this case the residual is 39.069 which is well below the threshold 56.94. So the attack remains hidden in the BDD module. In the next test, we inject random attack vector with the original measurements. For this case, the estimated measurements does not follow either the original measurements or the attacked ones, as shown in Fig. 2(d). The residual is also 4.3^5 which is noticeably large than the threshold. Therefore, the random attack is easily detected by the BDD module.

Now we demonstrate the data-driven attack based on measurement signals only. Here, 500 observations or measurement samples are recorded to create a measurement matrix assuming that the system states are following normal distribution with the same mean of the original measurement vector $\mathbf{z}$ [14]. Measurement noise is considered to be between $20 \sim 35$ dB. Based on the principles of PCA based blind attack, we design attack vector following Sect. 4 and inject with the original measurement. Similar to PCA [26], SVD based technique of [14] can also be used to construct data-driven attacks which does not need system Jacobian $\mathbf{H}$. For both of the cases the attacked measurement are different from the original measurements due to false injection (see Fig. 2(e) and (f)). However, for both of the cases, the estimated measurements and attacked measurements are indistinguishable. Moreover, the residuals for both of the cases are same to the no attacked case ($residual = 39.069$). So, these attacks also remain hidden in the BDD module. However, measurement based data-driven attacks using PCA can not guarantee its stealthiness as PCA can not estimate the measurement subspace accurately under missing values. To demonstrate the fact, we randomly consider 1 % missing values to the measurement matrix and then follow the same procedure that was used to construct PCA based attack. This time, we notice that the estimated measurement does not follow either the observed (attacked) measurements or the original ones, as shown in Fig. 2(g).

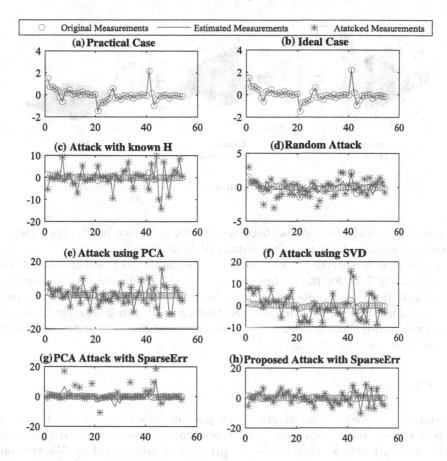

Fig. 2. Performance comparisons of different attack strategies (Color figure online)

The residual is also significantly large than the threshold which makes the attack detectable. As missing values in measurement data are very common in real-life smart grid operation, the data-driven stealthy FDI attack construction using PCA is completely challenging and not possible. Although it is a good news for the system operator, we strongly warn them that the intelligent attacker can still design stealthy attacks using sparse optimization. We show an example in Fig. 3, where there are only three missing values in the entire collection of the measurement matrix. Now, we employ the ALM based method discussed in the previous section which approximates the low-rank measurement matrix and the sparse missing value matrix (See Fig. 3). Now stealthy attack can be constructed using the approximated measurement matrix as demonstrated in Algorithm 1. To investigate the performance we again consider only 1 % missing values to the measurement matrix. Then the proposed attack strategy is employed. We observed that the estimated measurements follow the observed (attack)

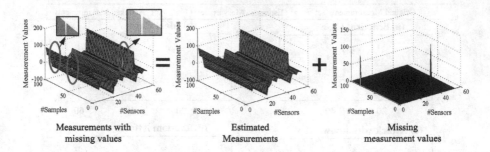

Measurements with missing values	Estimated Measurements	Missing measurement values

Fig. 3. The ALM method approximates the true measurement matrix

measurements and the residual becomes 39.069, as shown in Fig. 2(h). Hence, the attack remains hidden in the existing BDD module.

In our experiment, the attack construction using the proposed approach takes approximately 7.2 s for the 14 bus test system. All intelligent attacker can use more sophisticated computational technologies (e.g., high performance computing) to reduce the time requirements. As the system dynamics change gradually, any time requirements less than a few seconds to a minute is sufficient enough to construct stealthy attacks.

7   Conclusion

The paper discusses how an adversary can generate an attack vector by analysing system measurement data only. We demonstrate a much-improved version of state-of-the-art attacks that fail in the presence of missing values. We are concerned that the attack methodologies proposed in the state-of-the-art paper fails if there exists missing values in the measurement-which is a common occurrence in practical implementations. Hence, we demonstrate an improved attack that succeeds and remains hidden even in the presence of missing values. Consequently, the presented technique is much more practical. Furthermore, we show how the attack evades the State Estimator (SE) and Bad Data Detector (BDD) which are two key practical modules used in the real-world power system operation centres for measurement data processing. The impact of the proposed stealthy attack strategy is validated using these (e.g., SE and BDD) application modules considering IEEE benchmark test systems. The detection strategy for such attacks will be similar to the detection methods for FDI attacks proposed in the smart grid literatures.

References

1. Power systems test case archive. https://www.ee.washington.edu/research/pstca
2. Abur, A., Expósito, A.: Power System State Estimation: Theory and Implementation. Power Engineering (Willis). CRC Press, Boca Raton (2004)

3. Anwar, A., Mahmood, A.: Cyber security of smart grid infrastructure. In: Pathan, A.-S.K. (ed.) The State of the Art in Intrusion Prevention and Detection, pp. 139–154. CRC Press, Taylor & Francis Group, Boca Raton, Florida (2014)
4. Anwar, A.: Vulnerabilities of smart grid state estimation against false data injection attack. In: Hossain, J., Mahmud, A. (eds.) Renewable Energy Integration. Green Energy and Technology, pp. 411–428. Springer, Singapore (2014)
5. Anwar, A., Mahmood, A.N.: Anomaly detection in electric network database of smart grid: graph matching approach. Electr. Power Syst. Res. **133**, 51–62 (2016)
6. Anwar, A., Mahmood, A.N., Ahmed, M.: False data injection attack targeting the LTC transformers to disrupt smart grid operation. In: Tian, J., Jing, J., Srivatsa, M. (eds.) International Conference on Security and Privacy in Communication Networks. Lecture Notes of the Institute for Computer Sciences, Social Informatics and Telecommunications Engineering, pp. 252–266. Springer International Publishing, Switzerland (2015)
7. Anwar, A., Mahmood, A.N., Tari, Z.: Identification of vulnerable node clusters against false data injection attack in an AMI based smart grid. Inf. Syst. **53**, 201–212 (2015). Elsevier
8. Bi, S., Zhang, Y.J.: Graphical methods for defense against false-data injection attacks on power system state estimation. IEEE Trans. Smart Grid **5**(3), 1216–1227 (2014)
9. Candès, E.J., Li, X., Ma, Y., Wright, J.: Robust principal component analysis? J. ACM **58**(3), 11:1–11:37 (2011)
10. Esmalifalak, M., Nguyen, H., Zheng, R., Han, Z.: Stealth false data injection using independent component analysis in smart grid. In: International Conference on Smart Grid Communications, October 2011
11. Hug, G., Giampapa, J.: Vulnerability assessment of ac state estimation with respect to false data injection cyber-attacks. IEEE Trans. Smart Grid **3**(3), 1362–1370 (2012)
12. Jokar, P., Arianpoo, N., Leung, V.: Intrusion detection in advanced metering infrastructure based on consumption pattern. In: IEEE International Conference on Communications (ICC), June 2013
13. Kim, J., Tong, L., Thomas, R.: Data framing attack on state estimation. IEEE J. Sel. Areas Commun. **32**(7), 1460–1470 (2014)
14. Kim, J., Tong, L., Thomas, R.: Subspace methods for data attack on state estimation: a data driven approach. IEEE Trans. Sign. Process. **63**(5), 1102–1114 (2015)
15. Kosut, O., Jia, L., Thomas, R., Tong, L.: Malicious data attacks on smart grid state estimation: attack strategies and countermeasures. In: International Conference on Smart Grid Communications, October 2010
16. Lin, Z., Chen, M., Ma, Y.: The augmented lagrange multiplier method for exact recovery of corrupted low-rank matrices. Technical report, UIUC Technical report UILU-ENG-09-2214 (2009)
17. Lin, Z., Chen, M., Ma, Y.: Fast convex optimization algorithms for exact recovery of a corrupted low-rank matrix. Technical report, UIUC Technical report UILU-ENG-09-2214 (2009)
18. Liu, L., Esmalifalak, M., Ding, Q., Emesih, V., Han, Z.: Detecting false data injection attacks on power grid by sparse optimization. IEEE Trans. Smart Grid **5**(2), 612–621 (2014)
19. Liu, Y., Ning, P., Reiter, M.K.: False data injection attacks against state estimation in electric power grids. In: Proceedings of the 16th ACM Conference on Computer and Communications Security, CCS 2009, pp. 21–32. ACM, New York (2009)

20. Liu, Y., Ning, P., Reiter, M.K.: False data injection attacks against state estimation in electric power grids. ACM Trans. Inf. Syst. Secur. **14**(1), 13:1–13:33 (2011)
21. Ozay, M., Esnaola, I., Vural, F., Kulkarni, S., Poor, H.: Sparse attack construction and state estimation in the smart grid: centralized and distributed models. IEEE J. Sel. Areas Commun. **31**(7), 1306–1318 (2013)
22. Queiroz, C., Mahmood, A., Tari, Z.: SCADASim a framework for building scada simulations. IEEE Trans. Smart Grid **2**(4), 589–597 (2011)
23. Rahman, M., Mohsenian-Rad, H.: False data injection attacks with incomplete information against smart power grids. In: IEEE Global Communications Conference (GLOBECOM), December 2012
24. Valenzuela, J., Wang, J., Bissinger, N.: Real-time intrusion detection in power system operations. IEEE Trans. Power Syst. **28**(2), 1052–1062 (2013)
25. Xie, L., Mo, Y., Sinopoli, B.: False data injection attacks in electricity markets. In: IEEE International Conference on Smart Grid Communications (SmartGrid-Comm), pp. 226–231, October 2010
26. Yu, Z.-H., Chin, W.-L.: Blind false data injection attack using pca approximation method in smart grid. IEEE Trans. Smart Grid **6**(3), 1219–1226 (2015)
27. Zimmerman, R., Murillo-Sanchez, C., Thomas, R.: MATPOWER: steady-state operations, planning, and analysis tools for power systems research and education. IEEE Trans. Power Syst. **26**(1), 12–19 (2011)

Information Security in Software Engineering, Analysis of Developers Communications About Security in Social Q&A Website

Shahab Bayati[1(✉)] and Marzieh Heidary[2]

[1] ISOM Department, Business School, The University of Auckland, Auckland, New Zealand
S.bayati@auckland.ac.nz
[2] IT Department, Spark, Auckland, New Zealand

Abstract. By the growth of Internet based applications, security becomes an important part of software application development. Software developers should apply security modules, frameworks and technologies on their applications to reduce the security risks, bugs and vulnerabilities. This paper focuses on data analysis on the software development social Q&A Website content around security to elaborate the current state and trend of security issues in software engineering. For this purpose Stack Overflow data as the largest Q&A is selected to analyze. A framework is proposed for data collection and analysis from Stack Overflow. The result of analysis is presented in different schematic and tabular views and a brief discussion on each result is illustrated.

Keywords: Software engineering · Stack Overflow · Information security · Social Q&A · Developers community

1 Introduction

Software engineering (SE) is an iterative socio-technical process. Variety of technologies, programming languages, frameworks and toolkits are applied by different software developers in different stages of software development process. A robust software engineering process should consider security as an important part of development [1]. Security in software development life cycle (SDLC) plays an important role in success and failure of the project [2, 3]. Studies on SE bugs show the high impact and priority of security issues in SDLC. This research focuses on importance of security in software engineering by analyzing real data of software developers' communication.

Software engineers and developers use online forums and Q&A Websites to solve their problems [4]. Stack Overflow (SO) is the most famous online communities for software developers to ask their questions and searching for solutions from answered questions. This paper analyzed SO data about information security to know how software developers think about security. By investigation on SO data about security we can understand the real trends about security matters in SE process. This research clarifies who asked more questions related to security; which programming languages are mostly consider security issues; which technologies are highly impacted by security

© Springer International Publishing Switzerland 2016
M. Chau et al. (Eds.): PAISI 2016, LNCS 9650, pp. 193–202, 2016.
DOI: 10.1007/978-3-319-31863-9_14

questions in software engineering. Also it shows the importance of security in software development communications.

Stack Overflow with around 5 Million users by the time of writing this paper and more than 10 Million questions and Alexa world ranking less than 50 is the valuable resource for data analysis around development process. Along with Q&A facilities, it supports variety of features like user reputation management, up-voting and down-voting, badges and tags which helps to enrich the content of Q&A by social participation. SO as a great resource for data mining is considered in many MSR (Mining Software Repositories) researches [4–6]. Table 1 summarizes some basic statistics about SO. SO have valuable data about security in software engineering which is presented in this study.

Table 1. Basic statistics of Stack Overflow Website

Establish year	Users#	Questions#	Answers#	Comments#
2008	4,978,295	10,729,556	17,561,243	44,240,106
Tags#	Answered questions#	Votes#	Accepted answers#	Up-votes#
43,260	9,442,010	94,032,921	5,987,181	66,051,428

The main contribution of this research is showing the real trend of information security in software development based on the Q&A evidences on the largest development forum which is not presented in related previous works.

The rest of this study is structured as following. Next section presents related works. After that data collection, research questions and methodology are presented in Sect. 3. In Sect. 4 research results are illustrated. Discussion about results is presented in Sect. 5. Finally conclusion and future works are presented.

2 Related Works

In this section a brief review of previously published studies on the area of Stack Overflow mining and security in software engineering is presented.

2.1 Data Mining on Stack Overflow

A classification mining research is done on unanswered questions in SO Q&A Website. Normally the questions in SO are answered quickly (around 11 min after posting), but there are plenty of questions which remain unanswered. In mentioned research the main factors which may affect unanswered questions are gathered [6]. Reputation factors of SO users are mined in [7]. In a text analysis research on SO questions, authors used topic modeling analysis (LDA) to identify the relation among questions concepts, types and codes [8]. A paper focused on analyzing the expertise of developer who attends on SO to answer the questions. They checked the user participation in GitHub projects by technical term analysis and its relation to tags in SO [9]. Stack Overflow is mined to recommend the code example to the JQuery developers [4]. Mobile programming related posts on SO is mined by LDA text analysis approach in [10]. Mentioned studies shows the value of data analysis on SO.

2.2 Information Security in Software Engineering

In a text mining research on bug report categorization, Naive Bayes and TF-IDF are applied on Bugzilla reports. The main goal of this system was classification of reports into security bugs and non-security bugs. In the mention study TF-IDF approach performed better [11]. A dataset of high impact bugs are presented in a MSR research which listed security bugs as the high weight bugs in software development process. They manually tested bug reports of four open source projects. They categorized security bug more related to product bugs than process. In their views security bugs should be resolved with high priority. With the growth of Internet applications are security bugs received more priority [12].

In another MSR research characteristics of security bugs in Firefox project compare to other performance bugs. They mentioned that security bugs are more critical issues and should be treated faster. Their investigation shows that security bugs are assigned and fixed faster and also re-opened more frequently than others. Security bugs involve more files and developers [13].Importance of security consideration on Android application development to overcome SSL and MITM attacks is mentioned in [14].

Stack Overflow data is analyzed to find the role of security permission in Android applications. They found as the popularity of permission type grows the number of questions on SO also grows which results to better understanding in developers and less misuses [5]. Using online source code repositories like GitHub may lead to leakage of API secret keys in source codes which may happen to all collaborative development environments. A heuristic approach is used to detect and prevent this security problem in [15].

In a MSR research, the effect of pre-release security bugs on post-release vulnerabilities are discussed [16]. In another MSR research text mining applied on bug report for labeling them as security related or non security related. This is done by the reason of security bugs priority and mislabeled bugs [2]. All of these studies show the importance of security data analysis in software engineering.

3   Data Collection and Methodology

In this section data collection process from Stack Overflow social Q&A Website is introduced. SO publicly publishes its data in different formats. In this research SEDE (Stack Exchange Data Explorer) is used which provides an interface for running SQL queries. This dataset contains data about SO posts, comments, users, votes, badges, tags, feedbacks and etc. in a relational format. The result of queries can be downloaded in CSV format for further data analysis. SEDE originally is an open source project accessible from GitHub which uses Ms-Sql Server in data access layer. For the purpose of

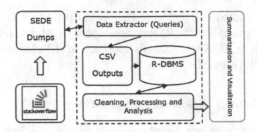

Fig. 1. Data collection and analysis framework.

this research the downloaded CSV files are cleaned and imported to R-DBMS for summarization and analysis. Figure 1 illustrates this process schematically.

Based on literature [17], to retrieve data about information security from SO we use SO security related tags and a dictionary of information security terms [18] along with an ontology of software security [19]. These collected terms are matched with similar SO tags in data extractor queries. Main terms in our queries include Security, Encryption, Authentication, SSL, Decryption, Sql-Injection, Script-injection, privacy, audit, policy, permission, access-control, hack, penetration, confidentiality and more other words which come out from this term list and correlate with them.

This research wants to answer the following questions on the gathered data from proposed abovementioned framework. 1. Which country users ask more about security issues of software development? 2. Which concepts in information security are asked more? 3. Which Programming Languages (PL), Operating Systems (OS), Mobile Technologies (MT) and Web Technologies (WT) have more questions about security? 4. How is the trend of information security related concepts in software engineering community? 5. How is different the answer rate of security related conversation in SE community with general questions? The answers of these questions can show the position of security in SDLC and the portion and priority of security concepts in SE. Based on our best of knowledge there is not any similar data analysis on software communities and repositories and this is the first work in this area which provide variety of future researches.

4   Research Results

This section provides the results of data analysis on different entities of SO. The first analysis aims to answer the question about which country developers are more active in security posts. To reach to this results top 500 users are gathered based on votes and activities, 121 users of top users do not fill their location positions and 10 users have inappropriate values for their locations. Many of the users just mentioned their states and city and countries together which are processed with a query to realize their main country. The pie charts in Fig. 2 show summary of this part of study. Table 2, shows more details. For data analysis and visualization we used R, Excel and D3.js.

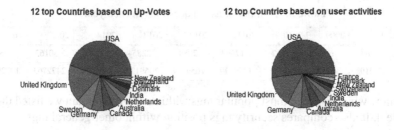

Fig. 2. Pie charts for the portion analysis of countries participation. Left up-votes, right activities

Table 2. Tabular view of each country participants.

		1	2	3	4	5	6	7	8	9	10
Up-votes	Country	US	UK	SW	DE	CA	AU	NL	IN	DK	PL
	Value	32551	12548	5752	4613	3642	3543	2482	2373	1688	1048
Activities	Country	US	UK	DE	CA	AU	NL	IN	SW	CH	NZ
	Value	350	104	46	37	32	24	23	13	13	12
Ratio	Country	SW	SK	PK	BG	ES	PL	DK	MT	IL	BD
	Value	442	283	222	188	157	150	141	136	137	122

Another area is about the response time to security questions. From the starting date of SO work in 2008 till December 2015 it is around 385 weeks and the average response time is about 61 min for security related posts which is about 20 min generally. In Fig. 3 the response time is presented over the time.

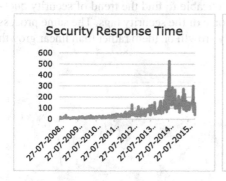

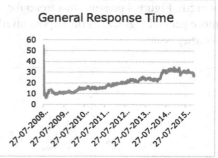

Fig. 3. Response Time (left: security related response time, right: general response time)

Amount of security related posts based on years is mentioned in Table 3. It shows growth of question amount year by year and compared to general questions. These amounts of post are added to previously asked questions.

Table 3. Number of asked questions in each year.

Year	2008	2009	2010	2011	2012	2013	2014	2015	Total
Security	1360	6612	11202	17663	20881	22854	23983	30163	135168
General	58219	343994	699773	1209317	1659879	2069137	2178788	2472763	10729556

To know which tags are most popular tags within security domain we listed them in the Table 4. It also compares security tags position within other general tags.

Table 4. Tags ranks by year.

	2008	2009	2010	2011	2012	2013	2014	2015	Total
C#	1	1	1	1	2	3	5	5	3
Java	3	2	2	2	1	2	2	2	2
Javascript	6	6	4	4	3	1	1	1	1
Android	587	66	10	5	5	5	4	3	5
PhP	7	4	3	3	4	4	3	4	4
.net	2	3	8	13	18	18	27	31	17
MySql	16	12	12	12	11	11	12	12	11
Html	12	13	13	11	9	7	7	8	8
Jquery	22	8	5	6	6	6	6	7	6
Security	35	52	58	77	102	125	151	152	102
Authentication	115	107	137	138	170	192	187	143	151
SSL	166	217	247	242	230	242	165	133	184

Based on available data on SEDE it is applicable to find the trend of security questions in SE. Figure 4 presents this trend for some of the security tags. The same process for more general tags shows the exponentially growth on them rather than linear growth on security issues.

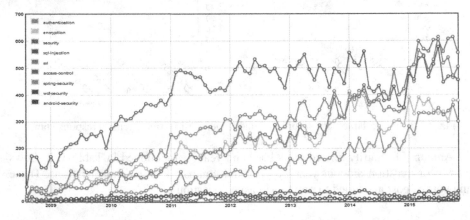

Fig. 4. Trends of some security related tags (Authentication, encryption, security, sql-injection, SSL, access control, spring-security, wcf-security and android-security)

Active users who answered to questions about security in SO are collected for a length of 7 years per week. Results are graphically shown in Fig. 5. As it is clear from this chart by the growth of SO members the active users in security area not only does not raise but also reduce. This problem should be investigated in future research.

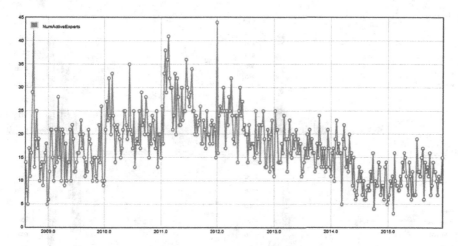

Fig. 5. Trend of active users for security questions

The maximum amount of tags for a post in SO is five. Security related tags can be used together in a post. Table 5 listed the number of times security tags co-occurred in SO posts with their frequencies. Only one question has five security tags together. Figure 6 shows this post.

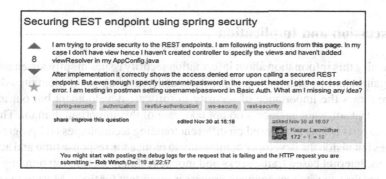

Fig. 6. The only question with five security related tags.

Table 5. co-occurred security tags. (only one question with five security tag)

Co-occurred	5	4	3	2
Post count	1	44	999	13431

Our next analysis counts for different software development technologies like NoSQL databases, Relational DBMSs, Web technologies and operating systems. The number of posts with the frequency of security related tags is presented in these charts. We have selected most trending items for each part. Figure 7 shows the details.

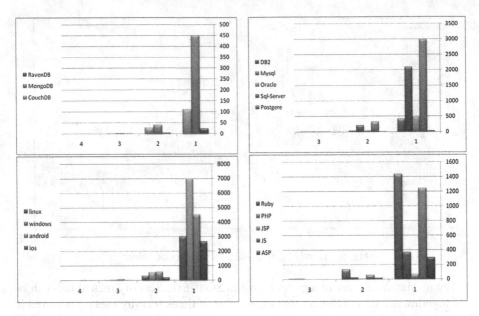

Fig. 7. Security related questions in different technologies, frameworks, OSs, DBMSs with frequency of security tags.

5 Discussion and Implication

By presenting this information about information security issues in software engineering we can gain valuable, actionable and practical understanding of this domain. Although literature shows the importance of security in software development but our analysis illustrated that software engineers do not take care of this important domain. The great need for experts in this area based on different trending technologies and programming languages felt inside the developer community to reduce the response time and accuracy of answers. Internet based frameworks and mobile technologies are top trending tags in security and the security management process is important for them. Most of the experts come from developed countries and big portion of them are from USA. It shows the lack of security efforts in other countries or may be lack of participation of other countries in Q&A knowledge sharing environments.

From response time analysis it can be argued that as we go further the response time to questions gets higher which may be affected by the quality of questions. Normally easy and general questions are asked in the beginning and more difficult ones are coming year by year. Another reason could be the lack of amount of active experts in the area and the growth in the number of questions on the other side.

6   Conclusion and Future Works

This study focuses on information security related posts and its surrounding data on the largest social Q&A community of software developers with different skills and expertise level. Stack Overflow data is used to analyze and response to research questions in this area. It shows that most of the security related questions are asked by US's developers and developed country users. Security, Authentication and SSL are the most usable security tags in questions. The response time for security questions are 3 times higher than the average. A trend analysis on most usable security tags on posts shows the linear growth of security questions compare to exponential growth in general questions amount. Also this study shows most trending technologies and frameworks have more questions about security.

For future works in this area, further analysis on security related data can be done. Furthermore, theoretical research studies can be applied on SO data to find which factors correlate with each other on some of the results shown in this study. Also expert reputation analysis in software security area is another research area for future. Moreover, text mining techniques like topic modeling and sentiment analysis can be applied in future on security related posts.

References

1. Van Wyk, K.R., McGraw, G.: Bridging the gap between software development and information security. IEEE Secur. Priv. **3**(5), 75–79 (2005)
2. Gegick, M., Rotella, P., Xie, T.: Identifying security bug reports via text mining: an industrial case study. In: 2010 7th IEEE Working Conference on Mining Software Repositories (MSR). IEEE (2010)
3. Tsipenyuk, K., Chess, B., McGraw, G.: Seven pernicious kingdoms: a taxonomy of software security errors. IEEE Secur. Priv. **3**(6), 81–84 (2005)
4. Zagalsky, A., Barzilay, O., Yehudai, A.: Example overflow: Using social media for code recommendation. In: Proceedings of the Third International Workshop on Recommendation Systems for Software Engineering. IEEE Press (2012)
5. Stevens, R., et al.: Asking for (and about) permissions used by android apps. In: Proceedings of the 10th Working Conference on Mining Software Repositories. IEEE Press (2013)
6. Asaduzzaman, M., et al.: Answering questions about unanswered questions of stack overflow. In: Proceedings of the 10th Working Conference on Mining Software Repositories. IEEE Press (2013)
7. Bosu, A., et al.: Building reputation in stackoverflow: an empirical investigation. In: Proceedings of the 10th Working Conference on Mining Software Repositories. IEEE Press (2013)
8. Allamanis, M., Sutton, C.: Why, when, and what: analyzing stack overflow questions by topic, type, and code. In: Proceedings of the 10th Working Conference on Mining Software Repositories. IEEE Press (2013)
9. Venkataramani, R., et al.: Discovery of technical expertise from open source code repositories. In: Proceedings of the 22nd International Conference on World Wide Web companion. International World Wide Web Conferences Steering Committee (2013)

10. Linares-Vásquez, M., Dit, B., Poshyvanyk, D.: An exploratory analysis of mobile development issues using stack overflow. In: Proceedings of the 10th Working Conference on Mining Software Repositories. IEEE Press (2013)

11. Behl, D., Handa, S., Arora. A.: A bug Mining tool to identify and analyze security bugs using Naive Bayes and TF-IDF. In: 2014 International Conference on Optimization, Reliabilty, and Information Technology (ICROIT). IEEE (2014)

12. Ohira, M., et al.: A dataset of high impact bugs: manually-classified issue reports (2014)

13. Zaman, S., Adams, B., Hassan, A.E.: Security versus performance bugs: a case study on firefox. In: Proceedings of the 8th Working Conference on Mining Software Repositories. ACM (2011)

14. Zhao, Y., et al.: A new strategy to defense against SSLStrip for Android. In: 2013 15th IEEE International Conference on Communication Technology (ICCT). IEEE (2013)

15. Sinha, V.S., et al.: Detecting and mitigating secret-key leaks in source code repositories. In: Proceedings of the 12th Working Conference on Mining Software Repositories. IEEE Press (2015)

16. Camilo, F., Meneely, A., Nagappan, M.: Do bugs foreshadow vulnerabilities? a study of the chromium project. In: 2015 IEEE/ACM 12th Working Conference on Mining Software Repositories (MSR). IEEE (2015)

17. Pletea, D., Vasilescu, B., Serebrenik, A.: Security and emotion: sentiment analysis of security discussions on GitHub. In: Proceedings of the 11th Working Conference on Mining Software Repositories. ACM (2014)

18. Kissel, R.: Glossary of key information security terms. DIANE Publishing, Collingdale (2011)

19. Raskin, V., et al.: Ontology in information security: a useful theoretical foundation and methodological tool. In: Proceedings of the 2001 workshop on New security paradigms. ACM (2011)

Author Index

Printed in the United States
By Bookmasters